Inorganic Materials Synthesis

ACS SYMPOSIUM SERIES **727**

Inorganic Materials Synthesis
New Directions for Advanced Materials

Charles H. Winter, EDITOR
Wayne State University

David M. Hoffman, EDITOR
University of Houston

American Chemical Society, Washington, DC

Library of Congress Cataloging-in-Publication Data

Inorganic materials synthesis : new directions for advanced materials / Charles H. Winter, David M. Hoffman, editors.

    p.   cm.—(ACS symposium series ; 727)

Includes bibliographical references and index.

ISBN 0–8412–3606–2

1. Solid state chemistry—Congresses. 2. Inorganic compounds—Congresses.

I. Winter, Charles H., 1959– . II. Hoffman, David M., 1955– . III. Series

QD478.I536   1999
541′.0421—dc21   99–21144
                                                       CIP

The paper used in this publication meets the minimum requirements of American National Standard for Information Sciences—Permanence of Paper for Printed Library Materials. ANSI Z39.48–1984.

PRINTED IN THE UNITED STATES OF AMERICA

Foreword

THE ACS SYMPOSIUM SERIES was first published in 1974 to provide a mechanism for publishing symposia quickly in book form. The purpose of the series is to publish timely, comprehensive books developed from ACS sponsored symposia based on current scientific research. Occasionally, books are developed from symposia sponsored by other organizations when the topic is of keen interest to the chemistry audience.

Before agreeing to publish a book, the proposed table of contents is reviewed for appropriate and comprehensive coverage and for interest to the audience. Some papers may be excluded in order to better focus the book; others may be added to provide comprehensiveness. When appropriate, overview or introductory chapters are added. Drafts of chapters are peer-reviewed prior to final acceptance or rejection, and manuscripts are prepared in camera-ready format.

As a rule, only original research papers and original review papers are included in the volumes. Verbatim reproductions of previously published papers are not accepted.

ACS BOOKS DEPARTMENT

Contents

SOLID-STATE INORGANIC CHEMISTRY DIRECTED TOWARD MATERIALS PROBLEMS

PREPARATION AND CHARACTERIZATION OF THIN FILMS

INDEXES

Preface

In recent years, there has been growing interest in materials chemistry, a field that focuses on the chemical aspects of material science. Chemists' interest in materials science has evolved with the realization that they are uniquely capable of finding improved syntheses of known materials and of synthesizing new ones, and that they can offer fundamental insight into materials properties. Chemists from all the broadly defined areas of chemistry, organic, inorganic, physical, and biological, have taken an interest in materials problems, and whole new subfields of study are emerging. For example, recent advances have led to distinct research areas emphasizing, for example, biomaterials, organic magnets, and organic light-emitting materials.

The subject of this book is inorganic materials chemistry, which is a subfield with roots in classical solid-state inorganic chemistry. Many of the key materials used, for example, in computer chips, efficient automobile engines, solar panels, fuel cells, batteries, magnetic resonance imaging, and many other modern devices, can be classified as inorganic materials. Despite the advances and the use of inorganic materials in real-world applications, however, a precise level of synthetic control in their preparations and a fundamental understanding of their physical properties are still lacking. The goal of this book is to document the latest advances in several important areas of inorganic materials chemistry, and through this process help define a field that is undergoing explosive growth.

The chapters in this book are based on a symposium entitled "New Directions in Materials Synthesis" that was held under the auspices of the Division of Industrial and Engineering Chemistry on September 7–11, 1997, at the 207th National Meeting of the American Chemical Society in Las Vegas, Nevada. The designed focus of the symposium was on studies involving materials synthesis rather than on measurement of properties. Such an emphasis was taken because it is most closely aligned with inorganic chemistry itself and, more importantly, the rate-limiting step in materials applications is often the synthesis of pure materials in a desired form (e.g., uniformly sized powders, conformal thin films, crystalline solids). The symposium served the valuable purpose of giving chemists with very different backgrounds, perspectives, and interests a forum to present their latest work and to learn about, discuss, and critique the work presented by others. The Division of Industrial and Engineering Chemistry was the ideal venue for the symposium because of its tradition of encouraging and facilitating the dissemination of interdisciplinary research results, especially those with real-world applications. A large contingent of young researchers, many of whom are only a few years into their independent research careers, were chosen to participate in the symposium and the writing of this book. By this approach, their enthusiasm and new perspectives were tapped, and their contributions can serve as a preview of emerging research.

The chapters are broken into two broad themes. The first emphasizes the application of solid-state inorganic chemistry to materials problems. Subjects covered include fundamental synthetic studies directed toward the identification of new inorganic phases with useful properties, the development of new porous inorganic phases that are designed to incorporate specific molecules and ions in a selective fashion, and the design of compounds with predictable magnetic properties. The chapters encompass applications in electronic materials, catalysis, sensors, ion exchange, the production and separation of enantiomerically enriched or pure organic compounds, the extraction of heavy-metal ions from aqueous waste streams, and molecular magnetic materials.

The second broad theme deals with the synthesis of thin films, which is further broken down into two subtopics. The first involves the use of metal–organic and organometallic compounds as precursors to inorganic thin films. Common threads in most of these chapters are the technique of chemical vapor deposition (CVD) and microelectronics applications. Specific topics covered include advanced barrier materials, copper thin films, and silicon–germanium–carbon alloys. Chapters describing fundamental studies that are directed toward the fabrication of metal–boride thin films by CVD techniques and the sol–gel synthesis of thin films relevant to the automotive industry are also included. The second subtopic focuses on the preparation of organic thin films by self-assembly of organic compounds on substrate surfaces. Topics include surface passivation, the creation of ordered structures on surfaces, and the preparation of surfaces containing desired functional groups.

The variety of subjects covered in this book reflects the diversity of perspectives brought to bear by chemists on materials chemistry problems. Practicing materials chemists and students of materials chemistry will use this book as a reference and gateway to the primary literature. Materials scientists and engineers, physicists, and other scientists from nonchemical disciplines will find it to be an excellent introduction to inorganic materials chemistry. Finally, chemical and materials educators who would like to incorporate the most recent research discoveries and directions in their classes should be able to use this book as a source of up-to-date information on inorganic materials chemistry.

CHARLES H. WINTER
Department of Chemistry
Wayne State University
Detroit, MI 48202

DAVID M. HOFFMAN
Department of Chemistry
University of Houston
Houston, TX 77204–5641

SOLID-STATE INORGANIC CHEMISTRY DIRECTED TOWARD MATERIALS PROBLEMS

Chapter 1

Synthesis and Characterization of Unusual Polar Intermetallics

Arnold M. Guloy, Zhihong Xu, and Joanna Goodey

Department of Chemistry and Texas Center for Superconductivity, University of Houston, Houston, TX 77204–5641

We have been exploring the synthesis and properties of new ternary and quaternary Zintl phases containing groups 13 and 14 post-transition elements. Interest on these materials are focused on understanding structure-bonding-property relationships in polar intermetallics that lie near the zintl border. It is anticipated that interesting electronic properties may be discovered in materials that lie between the normal intermetallics and zintl phases (semiconductors). Recent results from the systematic physical-chemical study on a number of anomalous Zintl phases such as KB_6 will be discussed. The synthesis and crystal structure of new intermetallic π-systems, $SrCa_2In_2Ge$ and $Ca_5In_9Sn_6$ will be discussed. The semiconducting phase, $SrCa_2In_2Ge$, features a $[In_2Ge]^{6-}$ allyl-like conjugated chain. The crystal structure and electronic structure of Ca_5In_9Sn features $[In_3]^{5-}$, an intermetallic cyclopropenium analog.

Introduction.

Exploratory Synthesis. Solid state chemistry, particularly the exploratory synthesis of inorganic materials, has experienced a renaissance over the past decade. This was brought about by the discovery of interesting and unprecedented properties such as high temperature superconductivity[1], charge density waves,[2] heavy fermions[3] and other unusual electronic and magnetic properties in novel solid state materials. More importantly, the potential of exploratory synthesis is emphasized by expectations that future new discoveries still await the "explorers". The importance of exploratory syntheses in the solid state sciences cannot be overemphasized, for what new properties could be measured and discovered without the synthesis of the solid state materials? At the least, exploratory synthesis and characterization contributes to building a larger body of knowledge that will add insight to the syntheses of other new solid materials. Once a new compound is synthesized, it is probable that some of its properties can be reasonably predicted. However, the cases in which the ability to predict properties fail are not unusual.[1-6]

The current approaches to the syntheses of new solid state materials can be classified into two general routes: (a) development of new synthetic methods, and (b) exploration of complex systems or "uncharted areas". The development of new synthetic routes to new materials is usually motivated by the desire to access new metastable or "kinetic" products which otherwise are unstable at high temperatures - the normal solid state synthetic condition. Hence, these new synthetic reactions normally occur at lower temperatures with the use of "solvents" such as molten salts, liquid metals or fluxes. Also in the last few years the development complex instrumentation has advanced the ability of solid state scientists to synthesize and characterize new thin film materials which are not accessible under normal solid state synthesis conditions.

The second general route to new materials deals with synthesis of complex compounds which may have four or more components. The syntheses of complex multi-component systems is normally accompanied by the unpredictability of the reactions and difficulty in characterization of non-homogeneous products. However, it is usually in this area that the most unusual and novel phases are observed. This is illustrated by the novel behavior of complex oxides such as the high-T_c superconducting cuprates and the complex manganates which exhibit colossal magnetoresistance (CMR).[7-8]

Intermetallic Compounds. The synthesis of complex intermetallic compounds offers a fertile ground for exploratory synthetic work. Not only does it offer a wide area for adventure in the many elements that are available, but also provides a testing ground for any new intuitive and innovative concepts of chemical bonding. Moreover, many of the novel physical (electronic and magnetic) phenomena have been discovered in many intermetallics. Among chemists, intermetallic compounds present particular problem: the seemingly unpredictable behavior of intermetallic phases based on current chemical concepts which is primarily derived from molecular chemistry.

The description of the chemistry and physics of intermetallics has evolved into classifications based on structure and bonding. A set of general rules in the formation of intermetallics was postulated by Hume-Rothery to explain the stability of all intermetallic structures.[9-10] These rules are based on the following: (1) electronegativity differences between the constituent atoms, (2) the tendency of s-p metals and transition metals to fill their s-p and d shells, (3) size-factor effects, (4) electron concentration, and (5) "orbital restrictions" - related to the symmetry conditions for orbital hybridization. A similar treatment by Pearson grouped intermetallics into three groups based on the most important factor that governs their crystal structure.[11] *Electron* compounds are governed by the number of electrons per atom, and an important class are Hume-Rothery phases. Laves phases and the so called Frank-Kasper phases form another group which rely on *geometric* requirements in their structural preference. The third class of intermetallics are called *valence* compounds. This group includes the normal valence compounds such as Mg_2Sn and compounds which obey the Zintl concept like NaPb. However, Parthe[12] and more recently, Girgis[13] showed that these three groups do not have clear boundaries and that in most intermetallic compounds all three principles, size, valence and electron concentration, are simultaneously involved. All of these factors must be recognized and exploited in any rational synthesis of new intermetallics. Hence, the difficult task of developing a rational approach in the synthesis of new intermetallic phases involves many interrelated and often complex factors.

4

Zintl Phases.

The Zintl concept provides an effective and useful way to rationalize the relationship between stoichiometry, crystal structure and chemical bonding in a large variety of main group intermetallics.[14] In this simple scheme, electropositive metals act merely as electron donors, donating electrons to their more electronegative partners. The latter elements then form bonds to satisfy their octets, yielding anionic clusters, chains, layers and networks. The application of the concept in rationalizing crystal structure-stoichiometry-property of tetrelides and pnictides has been quite successful. As a consequence the concept has been widely used in directing and rationalizing the syntheses of many main group post transition intermetallics.

The different binary phases of barium and silicon illustrate the application of the Zintl-Klemm concept. In BaSi the oxidation numbers are assigned as Ba^{2+} and Si^{2-}. Ba donates its two valence electrons giving Si a total of six valence electrons. The Zintl concept explains that Si^{2-}, isoelectronic to S (elemental S forms 2-bonded chains or rings) , forms two covalent bonds yielding zigzag chains of silicon. Similarly, BaSi2 yields Si^- which forms 3 homoatomic bonds to give tetrahedral $(Si_4)^{4-}$ clusters isostructural with white phosphorus. The analysis of Ba_3Si_4 is slightly more complicated because it assumes an Si_4^{6-} "butterfly" unit. This condition is in agreement with the octet rule since this polyatomic anion consists of two Si^{2-} (2-bonded) and two Si^{-1} (3-bonded) atoms. The Si_4^{6-} unit is arranged as a tetrahedron missing one edge due to the "reduction" of two Si atoms in $(Si_4)^{4-}$ clusters.

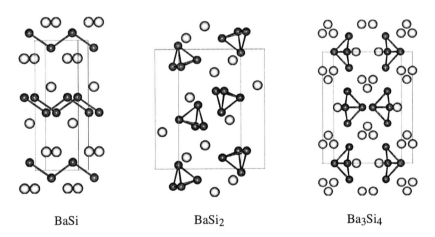

BaSi   BaSi₂   Ba₃Si₄

In the last few years, the Zintl concept has even been extended to rationalize the syntheses of many transition metal intermetallic phases.[15-17] The concept, in the framework of rigid band approach, has also been utilized to explain the chemical bonding in solids containing transition and main group metals. Zintl phases, with their various anionic frameworks, now form a large family of compounds whose electronic structures mostly conform to the 8-N rule. Basically they do not differ from insulators, semiconductors and semimetals and as a consequence, these phases have characteristics of normal valence compounds:[14,18]

The continuing evolution of the concept has led to explanations of non-stoichiometry, multi-center bonding and metal-metal bonding found in Zintl phases. Moreover, the assumption of the polar nature of Zintl phases has been validated by numerous theoretical and physical studies on Zintl and related polar intermetallic phases.[14,19-20] Modern band structure calculations performed on binary and ternary Zintl phases with the NaTl structure also confirm the conclusions drawn from a bonding scheme that is simultaneously ionic and covalent.[21-23] Altogether, quantum mechanical studies on Zintl phases underscore the brilliance and simplicity of the concept. The validity and chemical simplicity of the Zintl concept describes a smooth transition from molecular to solid state chemical concepts and allows the use of isoelectronic and electron-counting schemes in the bonding picture of intermetallics.

The effective charge transfer and the bonding character between atoms in intermetallics largely depend on differences in the electronegativities of the component elements.[24] Consequently, a criticism of the Zintl concept is the unreasonable assignment of ionic charges.[25] The description of Na^+Tl^- and Li^+In^- based on the Zintl concept is not contrary to the accurate theoretical and experimental picture if the assignment of "charges" were only in a *formal oxidation state* sense.[26] Therefore what is actually represented by the Zintl formalism is the number of occupied Wannier-like electronic states.[27,28] In Zintl phases, these states are mostly bonding or non-bonding, and are largely derived from the metalloid atomic states. The unoccupied antibonding states in normal Zintl phases and valence compounds, are mainly of the electropositive atom partner. It may then be assumed that if we could extend the electronic scheme to other intermetallic phases regardless of "class", we have the basis to probe into their chemical bonding and understand their physical properties. Nesper nicely illustrates this concept by describing electrons of intermetallics as being in "core potential valleys of two or more atoms, in two center two-electron bonds or multi-center bonds unaware that their core potential valleys may be covered by a sea of electrons high in energy ... Zintl phases and valence compounds correspond to the structured continents on earth, while in (inter)metallic phases the similarly well structured sub-ocean formations are shielded from direct view by the sea level (Fermi level)".[14]

Limits of the Zintl Concept. Despite the seemingly successful application of the Zintl concept, a number of violations are also observed. We could classify these anomalies into two general types: 1) those that seem to violate the structure-bonding relationships, with the number of bonds contradicting the number of available electrons; and 2) those with crystal structures that satisfy the Zintl rule but have "wrong" physical properties.

After careful investigations, many of the violations in the first group were later attributed to poor syntheses and characterization techniques. Many were also found to be impurity stabilized, or their stoichiometries were incorrectly assigned. However, there are a number of anomalies of the first kind which still remain unresolved and a thorough investigation of their physical and chemical behavior is necessary. The second group of violations represent chemical structures that are in agreement with the Zintl description but with physical properties that indicate some valence electrons actually behave as conduction electrons. This situation applies mostly in compounds that involve group 13 elements (trelides), where up to 5 electrons per atom might accommodated to satisfy the Zintl picture. It seems unlikely that these elements have enough effective core potential to accumulate high numbers of electrons without significant mixing with the electronic states of the metal component. In addition, group 13 elements and their compounds often do not conform to classical chemical bonding concepts. Hence, an empirical boundary between group 13 and 14 elements signifies the limit of the Zintl concept. Although the limit may be true in many cases, there are trelides like $LiMg_2Tl$ and $LiMgAl_2$ that still satisfy the requirements of being a Zintl phase.[29] Violations were also observed in polar intermetallics with group 14 and 15 elements and implies that the Zintl boundary is not absolute. In the spectrum from intermetallic compounds to Zintl or valence compounds to insulators, we observe a smooth transition in their chemical bonding going from metallic to ionic. At the border between Zintl phases and normal intermetallics, typical properties of Zintl phases diminish and metallic conductivity appears. Hence, near the Zintl limits, interesting and new structural chemistry of intermetallics is entirely possible.

Just how far to the left of the Zintl border will the main-group derived valence bands and the metal-derived conduction bands remain non-overlapping is yet to be determined. However, in complex intermetallics with late transition metals and many of the post-transition metals, we anticipate the charge transfer to occur from the most electropositive cationic partners to the covalent framework of transition and/or post transition metals. These complex intermetallics form what we term *polar intermetallics* and we expect the above valence generalities about Zintl phases to still apply.

General Synthesis and Characterization.

The synthesis of inorganic solids is often associated with considerable problems. Many are attributed to the high temperatures required to induce significant reaction rates and a sensitivity towards moisture, oxygen, or other adventitious impurities that complicate their handling. Furthermore, solid state syntheses suffer from the lack of purification methods. Incidental processes such as evaporation, reactivity of containers, and adventitious impurities affect the stoichiometry of the reactions in many unforeseeable ways. The limited number of characterization techniques is another important facet that requires attention. Nearly all effective characterization methods require an adequate degree of crystallinity, hence, effective means of crystal growth are often incorporated in any solid state preparation.

The completeness of a reaction sought - reaching a state of equilibrium - demands high temperatures because of small diffusion coefficients and large diffusion distances. As a rule, the products are what one expects from thermodynamic considerations. Since the kinetics of solid state reactions are mostly unknown and almost impossible to determine, other important aspects of chemistry on the mechanistic level and at lower temperatures are lost.

In the synthesis of polar intermetallics, one must be mindful of the following crucial points:

1) The lack of purification techniques during and after synthesis and the often perverse effects of impurities on high temperature synthesis requires the use of high purity starting materials (elements) and clean and inert containers (e.g. Ta, Nb, Mo). The use of silica and alumina containers is often avoided since their use have often led to erroneous results associated with the adventitious incorporation of oxygen or silicon.

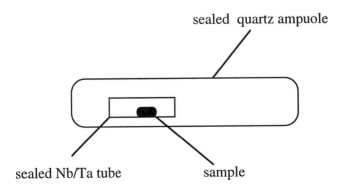

sealed quartz ampuole

sealed Nb/Ta tube   sample

2) By far the most satisfactory means of characterizing crystalline products is by X-ray diffraction. Hence, emphasis is given to obtaining and growing single crystals of unknown phases.

3) A new compound cannot be completely characterized until its structure, composition, and synthetic requirements are known. Low yield synthesis of a new solid phase is often a symptom of impurity stabilization or incomplete chemical analysis. The high yield synthesis of a new compound is a critical step of characterization.

4) The measurement of physical properties of new compounds and correlating their properties with crystal structure and chemical bonding is an integral part in any exploratory research.

Electron-Deficient Zintl Phases.

KB$_6$. We have made initial investigations on a number of previously reported phases which can considered anomalous Zintl phases. The results of our investigation on such a compound, KB$_6$ (CaB$_6$-type), illustrate the current problems concerning the validity of the Zintl concept..[30] The hexaboride KB$_6$ has been reported as a semiconductor which crystallizes with the cubic CaB$_6$ structure-type.[31] The structure as shown in Figure 1, can be described as a CsCl-type derivative wherein the metal is

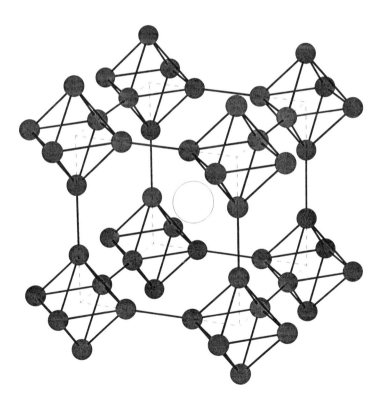

Figure 1. Crystal Structure of KB$_6$ - dark spheres: B (small), K (large); light spheres

located at the body center and the centers of the B_6 octahedral clusters are located at the corners of a cube. Furthermore, the vertices of the B_6 octahedra are bonded with the nearest neighboring cluster.

Compounds which crystallize with the CaB_6-type have been rationalized in terms of Wade's rules in that the network of corner-linked $[B_6]^{2-}$ octahedra: the 20-electron count derived from the stoichiometry results in 6 electrons used for exobonds to neighboring clusters and 14 electrons are used for the B_6 cluster bonds (2n+2). Hence, the alkaline-earth metal hexaborides have their valence requirements fully satisfied. In KB_6, the $[B_6]^-$ cluster network is apparently electron-deficient with only 19-electrons. Upon careful synthesis, structural and physical-chemical investigations, we find that the potassium hexaboride binary phase can be synthesized in high yield by the introduction of carbon - $KB_{6-x}C_x$ where $x = 0.8$-1.2. Its cubic structure features a network of corner-linked $(B_{6-x}C_x)^-$ octahedral clusters.

Subsequent electronic property measurements show that KB_5C is semiconducting and a normal Zintl phase. The presence of a homogeneity range in the composition with respect to carbon seems to defy the Zintl picture of the electronic structure. Electronic property measurements also show that the carbon-rich phases are metallic. This situation is analogous to the electronic picture of the electron-rich rare-earth metal hexaborides wherein the extra electrons are delocalized in the conduction bands. The carbon-poor phases are observed to be poor metals and their behavior follows those of heavily-doped semiconductors. These investigations on the interesting potassium boroncarbides point to the interesting possibilities in doping Zintl phases by introducing defects and nonstoichiometry. Other anomalous (electron-deficient) Zintl phases we are currently investigating are CaGaGe and CaInGe. [30]

New Zintl Phases with Inorganic π-Systems. In recent years, Corbett and coworkers, as well as Belin and coworkers, have reported on many interesting Ga, In and Tl cluster compounds.[18,32] The bonding in these 'electron-deficient' systems can be rationalized in terms of the usual Wade's rules as in the boranes, and most studies on trelides were spurred by the novelty of their cluster chemistry. Other reports on polar intermetallic trelides with mixed In-Ge anions have resulted in intriguing questions about the ability of In to accommodate high negative charges and open-shell electronic structures.[14,18,33] Our present studies on polar intermetallics involving trelide and tetrelide post-transition metals focus on the structural, chemical and physical characteristics of "electron-deficient" Zintl phases. These phases represent a class of Zintl phases where the normal picture of covalent networks of singly bonded metalloids may not be sufficient to satisfy the valence requirements of the anionic substructures. Our exploratory syntheses along the Zintl border have led to the discovery of such compounds - new complex Zintl phases, $SrCa_2In_2Ge$ and $Ca_5In_9Sn_6$.[34,35]

$SrCa_2In_2Ge$. This novel polar intermetallic crystallizes in the orthorhombic space group Pnna with 4 formula units in the unit cell.[34] The crystal structure of $SrCa_2In_2Ge$ is shown in Figure 2. The crystal structure is an ordered derivative of the CrB-type which can be described in terms of trigonal prisms of the metal atoms centered by metalloid/nonmetal atoms. This structure is common among alkaline and rare-earth metal monotetrelides. The main structural feature of the alkaline-earth and rare-earth metal monotetrelides are the parallel zigzag chains of 2-bonded tetrelide atoms making the alkaline-earth tetrelides normal Zintl phases and the rare-earth tetrelides as metallic phases. These observations result in a Zintl picture of their chemical bonding: $[(M^{2+}) + (2\text{-bonded})Ge^{2-}]$ in the alkaline-earth metal analogs; and $[(R^{3+}) + (2\text{-bonded})Ge^{2-} + 1e^-]$ in the rare-earth metal compounds.

In $SrCa_2In_2Ge$, the chain axis of the quaternary compound is about three times longer than that of the parent binary compound CaGe (CrB-type): $3c_{CaGe} \simeq b_{SrCa2In2Ge}$. The repeat unit in the non-metal atom chain, $[In_2Ge]^{6-}$, is [-In-In-Ge-In-In-Ge-] and the shortest distance between neighboring parallel chains is 4.6582(4) Å. The cations, Ca and Sr, are ordered in that only Ca atoms lie on the shared face between two In-centered trigonal prisms and Sr and Ca atoms lie on the shared face between Ge-centered and In-centered trigonal prisms. The resulting cation arrangement and nonmetal ordering result in short distances between adjacent In atoms (d= 2.772(2)Å) which are very short compared to those observed in the metal (d = 3.2-3.4Å), and other anionic In clusters and networks (d = 2.86 - 3.3Å).[29] The short In-In interatomic distances are comparable to those observed in cationic In fragments such as In_3^{5+}, In_5^{7+} and In_6^{8+} (d_{In-In} = 2.62 - 2.78Å) and in molecular In complexes (d = 2.81 - 2.93Å).[36] The calculated Pauling bond order (PBO) is about 2.3.[37] The observed Ge-In distances (d_{In-Ge} = 2.629(3)Å) are also comparably smaller than the sum of the single bond radii of In and Ge ($d1_{In-Ge}$ = 2.66 Å), PBO = 1.26.[37] The corresponding electron count per formula unit associated with the charge transfer from the cations does not satisfy the valence electronic requirement for a chain of singly bonded nonmetal atoms - ($[In_2Ge]^{8-}$).

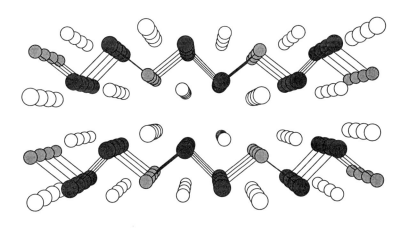

Figure 2. Crystal Structure of $SrCa_2In_2Ge$ - In, large dark spheres; Ge, small grey spheres; Sr, large white spheres; Ca, small white spheres.

Results of one-dimensional band structure calculations performed on the anionic $[In_2Ge]^{6-}$ show that all In-In bonding and non-bonding states are occupied and the In-Ge interactions are antibonding just below the Fermi level, E_f. The calculations also indicate reduced s-p hybridization as expected in heavier post-transition metals. Hence the lone pairs in indium are mainly 5p orbitals and the states that reflect In-In π-interactions lie below the Fermi level (E_f). Moreover, the In-In π*-antibonding states are unfilled and lie just above E_f with a small calculated band gap of @ 0.4 eV. Hence, the electronic structure and bonding in the compound can be represented by a conjugated $[In_2Ge]^{6-}$ π-system as illustrated in Scheme 1 with the first resonance structure as the more dominant one. The characteristics of the calculated band structure explains the short In-In and In-Ge interatomic distances observed in $SrCa_2In_2Ge$. Also, the electronic structure of the monomer $[In_2Ge]^{6-}$ is reminiscent of the simple organic allyl anion. Magnetic susceptibility measurements (M-T and M-H) indicate the compound is diamagnetic and this is consistent with being a semiconducting Zintl phase.[34]

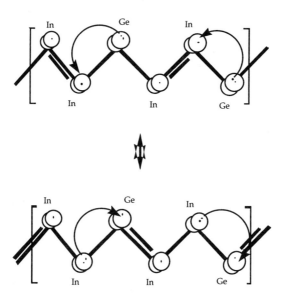

Scheme 1. Resonance structures of the conjugated $[In_2Ge]^{6-}$ chain

The discovery of this unprecedented compound strongly suggests the possibility of finding other unusual inorganic π-systems in "electron-deficient" Zintl phases. Also, the use of different sizes of isoelectronic cations, Ca and Sr, is crucial to the synthesis of this new complex Zintl phase and must be considered in the synthesis of other unusual Zintl phases. Introducing "electron deficiency" into robust intermetallic structures by cation or anion substitution may lead to "holes" in the valence band or to localized multiple bonds in these inorganic systems and these conditions might result in interesting electronic properties.

$Ca_5In_9Sn_6$. The interesting intermetallic phase $Ca_5In_9Sn_6$ crystallizes in a hexagonal structure in space group $P6_3/mmc$ with a = 6.7091(5)Å and c = 26.9485(9)Å.[35] The crystal structure, as shown in Figure 3, can be derived from two intermetallic structures, namely, cubic closed packed (ccp) $AuCu_3$-type and hexagonal closed packed (hcp) Ni_3Sn-type. The relationships between these structures are based on the differences in the stacking order of the intermetallic AX_3 closed packed layers. The crystal structure of $Ca_5In_9Sn_6$ is composed of mixed hcp

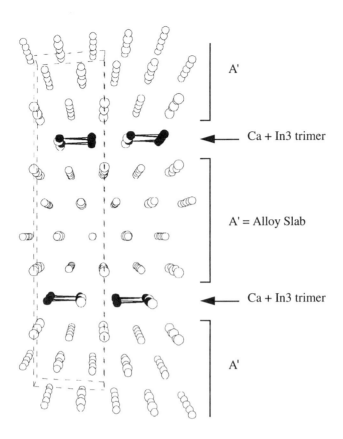

Figure 3. Crystal Structure of $Ca_5In_9Sn_6$. The alloy slab is composed of four $Ca(In,Sn)_3$ closed-packed layers. The In-In distance in the trimer is 3.0 Å, other In-In and In-Sn distances range from 3.25 - 3.4 Å.

and *ccp* layers of $Ca(In/Sn)_3$ in a ratio of 1:4. The unit cell contains 10 closed packed layers normal to the c-axis with *abcbacbabc* as the repeat stacking sequence. There are only two *hcp* layers in the unit cell and these are composed only of In and Ca atoms. The other closed packed layers (*ccp*) consists of Ca, In and Sn atoms. Within the *hcp* layers, indium atoms are displaced from the ideal closed packed positions resulting in the formation of In trimers. The In-In distances within the trimer is 3.045(3)Å, comparable to the Pauling In-In single bond distance.[37] The distances among In and Sn atoms within the *ccp* layers are relatively longer, 3.32-3.36Å and are comparable to those observed in non-Zintl intermetallic phases of In and Sn such as $LaIn_3$ and $CaSn_3$.[29]

Full 3-D band structure calculations show that the compound is essentially metallic with no observable gaps around the Fermi level.[35] Moreover, the contributions of the In trimers to the total density of states are small. Investigations on the more important question about the bonding within the In trimer show that all the bonding states of the In_3 trimer are occupied at the electron count corresponding to $Ca_5In_9Sn_6$. Separate molecular orbital calculations on an isostructural In_3 trimer shows an excellent correlation between the solid state band structure calculation (fragment molecular orbital analysis) and the In_3 molecular orbital scheme. This results in the unusual formulation of the In trimer as $[In_3]^{5-}$ anion with a 14 electron count. This electronic scheme for In_3 trimers is substantially different from those observed in cluster network compounds wherein In triangles are formally analogous with their arachno-B_3H_9 derivatives.[38] The band structure calculations also indicate that the rest (majority) of the occupied states in the DOS are expectedly derived from the delocalized states in the *ccp* intermetallic slabs.

The $[In_3]^{5-}$ trimers in $Ca_5In_9Sn_9$ are isoelectronic and analogous with cyclopropenium, $C_3H_3^+$, the simplest aromatic hydrocarbon. Furthermore, the surprisingly high formal charge on the indium trimer indicated by the calculations have also been observed in other In/Ge compounds such as La_3In_4Ge, La_3InGe and La_3In_5.[33] Physical property measurements on powdered samples of $Ca_5In_9Sn_6$ show it to be a typical intermetallic phase, i.e., metallic and Pauli-paramagnetic, with no superconducting transitions above 4K.[35]

The unprecedented "Zintl-alloy" phase resides well within the Zintl border. Its crystal and electronic structure can be rationalized in terms of two components: the "normal" intermetallic slab and a "molecular" or Zintl layer. This picture agrees well with the structural characterization, property measurements and band structure calculations. The results of these measurements indicate that the intermetallic (*ccp*) slabs dominate the electronic structure and properties of the compound. However, submerged underneath the sea of conduction electrons or delocalized states due to the intermetallic slabs lie the valence or localized states of the $[In_3]^{5-}$ trimers. Hence, it appears that for compounds that lie on the Zintl border the interplay between the structural factors of delocalized intermetallic and localized Zintl constituents, as well as, the nature of the interactions between them significantly determine their electronic structures and physical properties. The idea of tuning the interaction between the seemingly opposing factors provides a new concept to tuning properties of solid state materials.

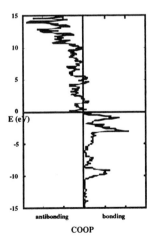
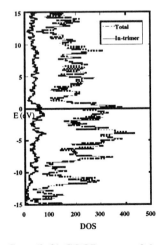

Figure 4. COOP and DOS curves for $Ca_5In_9Sn_6$: (left) COOP curve of the In-In interaction in In-trimer; (right) DOS curves, the dotted line is total DOS; solid line, In-trimer.

Implications.

The syntheses of $KB_{6-x}C_x$, $SrCa_2In_2Ge$ and $Ca_5In_9Sn_6$ nicely illustrate the strong possibility of finding new and unusual intermetallic analogs of unsaturated and/or aromatic hydrocarbons in the border between normal intermetallic compounds and semiconducting Zintl phases. The discovery of these unprecedented compounds mirrors the behavior of hydrocarbons and boranes in that there are two ways these compounds accommodate electron-deficiency, namely, the formation of multiple bonds and 3-center-2-electron bonds. Hence, these compounds are also expected to exhibit unique crystal structures and electronic properties. Efforts to explore this new research area must include careful synthesis, as well as, complete structural, chemical and physical characterization. Equally important is the application of effective theoretical bonding concepts in extracting useful chemical bonding descriptions from complex band structures of these unusual solid state materials.

Acknowledgements. We thank J. Korp for assistance with crystallography. We also thank Professor T.A. Albright for helpful discussions and use of the Extended Hückel Programs. Funding from the Petroleum Research Fund and the State of Texas through the Texas Center for Superconductivity is gratefully acknowledged. A.M.G. acknowledges a CAREER award from the National Science Foundation.

References

1. Bednorz, J.G; Muller,K.A. *Z. Phys.* **1986** *B64*, 189.
2. Wilson, J.A.; Di Salvo, F.J.; Mahajan, S. *Phys. Rev. Lett.* **1974** *32*, 882.
3. Sleglich, F.; Aarb, J.; Bredl, C.D.; Lieke, W.; Meschede, D.; Franz, W.; Schäfer, H. *Phys. Rev. Lett.* **1979** *43*, 1892.
4. Di Salvo, F.J. In *Advancing Materials Research*; Psaras, P.A. and Langford, H.D., Eds.; National Academy Press: Washington,D.C., **1987,** pp 161-175.
5. Michel, C.; Raveau, B. *Rev. Chim. Miner.* **1984**, *21*, 407.
6. Bednorz, J.G.; Muller, K.A. *Science* **1987**, *237*, 1133.
7. Baibich, M.; Broto, J.; Fert, A.; Nguyen van Dau F.; Petroff, F., *Phys. Rev. Lett.* **1988**, *61*, 2472.
8. Parkin, S; *Annu. Rev. Mater. Sci.*. **1995**, *25*, 357.
9. Hume-Rothery, W. *J. Inst. Metals* **1926**, *35*, 307.
10. Westgren, A.; Phragmen, G. *Arkiv. Mat. Astron. Fysik.* **1926**, *l9B*, 1.
11. Pearson, W.B. *The Crystal Chemistry and Physics of Metals and Alloys*; Wiley-Interscience: New York, 1972, Chapter 5.
12. Parthe, E. *Z. Krist.* **1961**, *115*, 52.
13. Girgis, K. In *Physical Metallurgy*; Cahn, R.W. and Haasen, P.; Eds.; North-Holland: Amsterdam, 1983, Vol. l, Chapter 5, pp 220-269.
14. a) Nesper, R. *Prog. Solid State Chem.* **1990**, *20,* 1; b) Nesper, R. *Angew. Chem. Int. Ed. Engl.* **1991**, *30*, 789.
15. Schuster, H.U. *Nova Acta Leopoldina* **1985**, *59,* Nr. 264, 199.
16. Jeitschko, W.; Reehuis, M. *J. Phys. Chem. Solids* **1987**, *48,* 667.
17. Kauzlarich, S. *Comments Inorg. Chem.* **1990**, *10*, 75. and *Chemistry, Structure and Bonding of Zintl Phases and Ions,* Kauzlarich, S.; Ed.; VCH Publishers: New York, 1996.
18. Corbett, J.D. In *Chemistry, Structure and Bonding of Zintl Phases and Ions,* Kauzlarich,S; Ed.; VCH Publishers: New York, 1996; and references therein.
19. Novotny, H.; Reichel, H. *Powder Metall. Bull.* **1956**, *7*, 130.

20. McNeil, M.B.; Pearson,W.B.; Bennett, L.H.; Watson, R.E. *J. Phvs.C.* **1973**, *6,* 1.
21. Christiansen, N.E. *Phys. Rev.* **1985**, *B32,* 207.
22. Christiansen, N.E. *Phvs. Rev.* **1985**, *B32,* 6490.
23. Schmidt, P.C. *Structure and Bonding* **1987**, *65,* 91.
24. Watson, R.E.; Bennett, L.H. In *Charge Transfer/Electronic Structure of Alloys;* Bennett, L.H., and Willens, R.H., Eds.; Metallurgical Society of AIME: New York, 1974, pp 1-25.
25. Bennet, L.H. In *Developments in the Structural Chemistry of Allov Phases;* Giessen, B.C., Ed.; Plenum Press: New York, 1969, pp. 41 - 44.
26. Sleight, A.W. *Proceedings of the Robert A. Welch Foundation Conference on Chemical Research* 1988, *XXXII,* Chapter 6.
27. Anderson, P.W. *Proceedings of the Robert A. Welch Foundation Conference on Chemical Research* l988, *XXXII,* Chapter 1, p 1.
28. Rudnick, J.; Stern, E.A. *Phvs. Rev.* **1973**, *B7,* 5062.
29. a) Villars, P.; Calvert, L.D. *Pearson's Handbook of Crvstallographic Data for Intermetallic Phases;* American Society for Metals: Metals Park, OH, 1985; 3 vols.; b) Pearson, W.B. *The Crystal Chemistry and Physics of Metals and Alloys;* Wiley-Interscience: New York. 1972.
30. Xu, Z.; Kim, M. and A.M.Guloy, unpublished work.
31. Perkins, P.G.; Sweeney, A.V.J. *J. of Less-Common Metals* **1976**,*47,* 165.
32. Belin, C; Tillard-Charbonnel, M. *Prog. Solid State Chem.* **1993**, *22,* 59.
33. Guloy, A.M.; Corbett, J.D. *Inorg. Chem.* **1996**, *35,* 2616.
34. Xu, Z.; Guloy, A.M. *J. Am. Chem. Soc.* **1997**, *119,* 10541.
35. Xu, Z.; Guloy, A.M. *J. Am. Chem. Soc.* in press.
36. Downs, A.J. In *Chemistry of Aluminium, Gallium, Indium and Thallium;* Downs, A.J., Ed.; Blackie Academic and Professional: London, 1993; pp. 65-70
37. Pauling, L. *The Nature of the Chemical Bond;* Cornell University: Ithaca, 1960.
38. Blase, W.; Cordier, G.; Vogt, T.; *Z. Anorg. Allg. Chem.* **1991**, *606,* 79.

Chapter 2

Exploitation of Zintl Phases in the Pursuit of Novel Magnetic and Electronic Materials

Susan M. Kauzlarich, Julia Y. Chan, and Boyd R. Taylor

Department of Chemistry, University of California, One Shields Avenue, Davis, CA 95616

A brief historical overview on Zintl phases is provided and an introduction to the Zintl concept is given. This group's strategies for the synthesis of new solid phases and new forms of materials prepared by means of solution methods are discussed. An approach to synthesis of transition metal Zintl phases is outlined and our discovery of colossal magnetoresistance (CMR) in $Eu_{14}MnSb_{11}$ is presented. Also, the use of Zintl phases to produce nanocrystals of Si and Ge is presented. FTIR spectroscopy of Ge nanocrystals provides evidence that these nanocrystals can be terminated with alkyl groups.

Eduard Zintl lived from 1898-1941, a relatively short time, considering his impact on solid state and inorganic chemistry.[1] He was a German scientist who started his career at Freiburg University and went on to the Institute for Inorganic Chemisty at the Technische Hoschchule Darmstadt. One of his interests was in the synthesis and characterization of binary intermetallic compounds composed of alkali metals and main group elements. He was able to characterize these solid phases by x-ray powder diffraction and developed methods for handling air-sensitive compounds. By determining the crystal structure of these phases, he could distinguish the phases as either salt-like or metallic type structures. Since many of these phases were considered to be intermetallics, the idea that one could understand the structures as being composed of cations and anions or polyanions which obey valence rules was a significant contribution to our understanding of the bonding. In addition, he was also interested in the ammonia solution chemistry of these alkali metal main group phases and speculated, based on potentiometric titration experiments, on the existence of polyanions such as As_3^{3-}, As_5^{3-}, As_7^{3-}, Sn_9^{4-}, and Pb_7^{4-}. These polyanions have since been isolated and their structures confirmed by single crystal x-ray diffraction.[2]

After Zintl's death, F. Laves proposed calling those intermetallic compounds which had been described by Zintl and are understood by valence rules Zintl Phases. Unlike the intermetallic phases, these phases crystallize in isotypic salt-like structure types and are considered to be semiconductors. This classification has since been enlarged to include more complex phases than simple binaries. Generally, one can consider phases such as A_xB_y and $A_xM_yB_z$ being composed of an electropositive element A which donates its electrons to the more electronegative component, B_y or M_yB_z, which uses the electrons to satisfy valence. This results in structures with cations and isolated anions, polyatomic anions, networks, layers, or chains. It is the documentation of the remarkable structures of Zintl phases published by Schäfer which attracted our attention to this area of chemistry.[3-5]

We are interested in both the structure of Zintl phases and their reactivity in solution. We utilize the structural principles of Zintl phases to propose and synthesize new transition metal Zintl phases. In addition, we are interested in using Zintl phases to prepare novel phases of well known materials. The large negative charge of the polyanions found in some Zintl phases makes them useful as synthetic reagents. Fine magnetic particles have been prepared by reacting metal halides with Zintl phases.[6] We have utilized the Zintl phases ASi and AGe to prepare Si and Ge nanoclusters (A = alkaline metal). These phases contain either Si_4^{4-} or Ge_4^{4-} polyhedra. This contribution contains a brief introduction to our work on transition metal Zintl phases and the use of Zintl phases to produce group 14 nanoclusters.

Transition Metal Zintl Phases.

Our interests in structure and magnetism led us to the idea of trying to synthesize transition metal analogs of Zintl phases. We were intrigued by the fact that many of the Zintl phases make a transition from semiconducting to metallic behavior: Zintl compounds are postulated to be semiconductors, but by changing the identity of the anion (i.e. As by Sb or Sb by Bi) there is a transition to metallic behavior. If the transition metals with partially filled d orbitals are introduced into a Zintl structure, unusual magnetic and electronic properties may result because of this proximity to a metal-insulator border. The Zintl concept helps to target compounds of particular structure types and the phases that might be expected to have unusual properties are those that are referred to as "polar intermetallic". That is, they have some aspects of covalent as well as metallic bonding. Therefore, we are utilizing the well-known principle of looking for novel magnetic and electronic effects at a metal-insulator border.

The first transition metal analog of the $Ca_{14}AlSb_{11}$ structure type, discovered by Cordier et al,[7] was synthesized by this group in 1989.[8] Since that time, a large number of new compounds have been prepared.[9-13] This structure type fits into the large class of compounds referred to as Zintl compounds. The electrons donated from the alkaline earth atoms are used to satisfy valence for the main group atoms. For this structure type, one formula unit consists of 14 A^{2+} cations, 4 Pn^{3-} anions, a MPn_4^{9-} tetrahedron, and a Pn_3^{7-} linear anion (Pn = P, As, Sb, Bi).

The Pn atoms in this structure are considered to be anions. The Pn_3^{7-} anion can be recognized as a 22 electron species, isoelectronic with I_3^-. This Pn_3^{7-} unit is a linear anion and can be considered a hypervalent 3-center-4-electron species. *Ab initio* calculations as well as IR and Raman spectroscopy are consistent with this interpretation of the structure.[11,14]

The family of compounds that crystallize in this structure type has been enlarged to include the metals, Mn, Nb, Zn, and Cd.[9,12,15] The tetragonal Nb phases, $A_{13}NbSb_{11}$ (Sr, Eu) are defect variants of the $Ca_{14}AlSb_{11}$ structure with fractional cation vacancies.[9]

The $A_{14}MnPn_{11}$ compounds have been shown to have unusual magnetic and electronic properties.[12] In these transition metal Zintl compounds, the metal in the tetrahedron is formally Mn^{III}, a d^4 ion. Temperature dependent magnetic susceptibility of the arsenic analogs is consistent with this assignment and the compounds show paramagnetic behavior that can be fit by the Curie-Weiss law. The heavier analogs, Pn

= Sb, Bi, show spontaneous magnetization with $A_{14}MnSb_{11}$ (A = Ca, Sr, Ba) and $A_{14}MnBi_{11}$ (A = Ca, Sr) showing ferromagnetic behavior, and $Ba_{14}MnBi_{11}$ showing antiferromagnetic behavior. Temperature dependent resistivity measurements as well as the magnitude of the resistivity suggest that the Sb compounds are semimetals and that the Bi compounds are barely metallic. The magnetic coupling has been attributed to indirect exchange of the localized moments on the Mn mediated by conduction electrons according to Rudderman-Kittel-Kasuya-Yosida (RKKY)[16,17] interaction.[18-20] The highest temperature for magnetic ordering of the alkaline earth analogs is 65 K for $Ca_{14}MnSb_{11}$.[20]

$Eu_{14}MSb_{11}$ (M = Mn, In). An extension of this work to the divalent rare earth analogs was undertaken with the expectation that the addition of an f electron ion will contribute to interesting magnetic and electronic properties. Only one example will be provided in this chapter, that of $Eu_{14}MSb_{11}$ (M = Mn, In).

The $Eu_{14}MSb_{11}$ can be prepared by reacting stoichiometric amounts of the elements in a sealed tantalum tube which is further sealed in a fused silica tube under 1/5 atmosphere purified argon. High yields of polycrystalline pieces were obtained by heating the mixtures to temperatures at 1000 °C for periods between 24 hours to 5 days and cooling the reaction to room temperature at rates of 5 to 60 °/hr. The best method to date for producing crystal from the elements were in a 2-zone furnace with T_{high} = 950 °C and T_{low} = 1000 °C for 7 days. The samples are characterized by powder and single crystal x-ray diffraction.

Eu^{2+} and Sr^{2+} are almost the same size, so one might assume that the Eu containing compounds might have similar ordering temperature as the Sr compounds. However, $Sr_{14}MnSb_{11}$ shows a ferromagnetic transition at 45 K[20] and $Eu_{14}MnSb_{11}$ shows a ferromagnetic transition at 92 K (See Figure 1).[13] The substitution of Eu for Sr doubles the transition temperature. The high temperature data behave according to the Curie-Weiss Law and the inset shows the inverse susceptibility as a function of temperature. It is linear and when fit to the equation, $\chi = \chi_0 + C/(T-\theta)$, one obtains χ_0 = 0.09(0) emu/mole, C = 88.9(1), and θ = 93.55(3) K. The value of θ is in good agreement with the observed Curie T of 92K. The moment is obtained from the equation, $\mu_{eff} = (8C)^{1/2}$ and gives the value of 27.0(1) μ_B. This is lower than what one would expect based on 14 Eu^{2+} ions and $1Mn^{3+}$ ion. The low temperature magnetic susceptibility data show a difference in the zero-field cooled (ZFC) and field cooled (FC) behavior and there is a peak at 15 K. A hysteresis curve is the measurement of the magnetic moment as a function of field. Typically, the magnetic field is swept positive, reversed to zero and negative, and then reversed again to positive. If the magnetization saturates, the total moment for the system can be deduced. Analysis of the hysteresis loop for Eu provides a saturation moment of 102 μ_B, in good agreement with what is expected for 14 Eu^{2+} ions and 1 Mn^{3+} ion.

The transition temperature can be attributed to ordering of the Mn spins. Figure 2 shows the magnetic susceptibility of $Eu_{14}InSb_{11}$ and there is clearly no ferromagnetic ordering. However, there is a low temperature transition at approximately the same temperature as that observed in the Mn compound which can be attibuted to magnetic ordering of the Eu^{2+} ions.

We were quite surprised, however, when we measured the resistivity of these compounds. There was a pronounced transition in the resistivity of the Mn containing compound which coincides with the magnetic transition. $Eu_{14}InSb_{11}$ shows semiconducting behavior.

Magnetoresistance (MR) is the change in resistance as a function of magnetic field. The magnetoresistance ratio is defined as MR = $[(\rho(H=0)-\rho(H)/\rho(H=0))$ x 100 %]. Whereas many magnetic materials show some MR, it is typically positive and less than 2%. There has been a great deal of interest in negative magnetoresistance because of

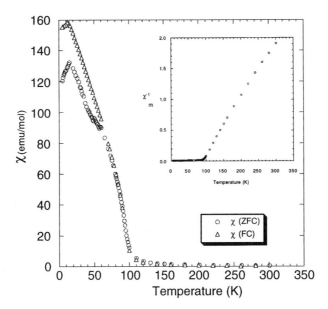

Figure 1. Temperature dependent magnetic susceptibility of $Eu_{14}MnSb_{11}$. Zero-field cooled (ZFC) and field cooled (FC) data taken at 1000 Gauss. Inverse susceptibility versus temperature is shown as an inset. (Reproduced with permission from ref. 13, Copyright 1997 ACS)

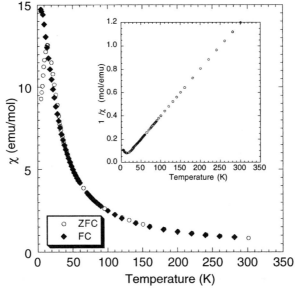

Figure 2. Temperature dependent magnetic susceptibility of $Eu_{14}InSb_{11}$. ZFC and FC data taken at 1000 Gauss. Inverse susceptibility versus temperature is shown as an inset. (Reproduced with permission from ref. 13, Copyright 1997 ACS)

the possible application of this type of material in read head devices. The term "giant magnetoresistance" or GMR was used to describe magnetic multilayers and granular materials which show a large (typically >5%) negative MR ratio.[21] The discovery of "colossal" magnetoresistance in the rare-earth perovskites, $Ln_{1-x}A_xMnO_3$ (Ln = lanthanide, A = Ca, Sr, Ba, Pb) has stimulated further research in this field. Many of these perovskites show negative MR ratios of 100%. These compounds were initially called colossal to indicate the size of the MR ratio and to distinguish them from the better studied multilayered materials. It has been suggested that CMR materials have MR effects that are associated with the magnetic transition and that the MR does not saturate below fields of 6 Tesla.[22]

The similarity of the resistance of $Eu_{14}MnSb_{11}$ to the perovskites CMR materials prompted our measurements of MR in $Eu_{14}MnSb_{11}$. In order to measure magnetoresistance, four Pt leads were attached to a single crystal of $Eu_{14}MnSb_{11}$. Figure 3 shows a view of the crystal with leads attached. The outer leads are for the current and the inner two leads are for measuring the voltage drop. The long axis of the crystal is the c axis and it can be oriented parallel or perpendicular to the magnetic field. Figure 4 shows the resistivity of $Eu_{14}MnSb_{11}$ as a function of magnetic field. Similar to the perovskite phases, there is a suppresion of the resistivity as a function of increasing magnetic field. There are two temperatures at which this effect is maximized, 92 K and 5 K. The low temperature effect is attributed to ordering of the Eu^{2+} moments and the high temperature effect to ferromagnetic ordering of the Mn^{3+} moments. The low temperature effect can be removed by introducing a nonmagnetic Ca^{2+} cation to replace one of the Eu^{2+} cations (see Figure 5). This substitution has little effect on the transition temperature, which is expected since the high temperature transition is attributed to the ordering of Mn moments.

Figure 6 shows the resistivity as a function of magnetic field for $Eu_{14}MnSb_{11}$. Above the transtion temperature of 92 K, the curves are convex, indicating that the resistivity is saturated. At and below the magnetic transition, the curves are quite concave, indicating that this phase will not saturate until the field is significantly larger than 5 Tesla.

The fact that the MR effect is associated with the onset of ferromagnetic behavior and the resistance does not saturate at fields less than 5 Tesla classify this compound as a CMR material. We are currently investigating the synthesis of the P and As analogs of this structure type. We are continuing to investigate the synthesis of other transition metal Zintl phases. New transition metal Zintl phases will be found and with them, new and unusual properties.

Group 14 Semiconductor Nanoclusters.

Since it was first suggested that the visible luminescence seen in porous silicon could result from quantum confinement,[23] interest in making nanosized silicon particles has increased greatly.[24-29] How nano-sized particles of silicon and germanium luminesce despite their indirect band gap is not yet fully understood but it is generally agreed that quantum confinement is involved. This is the most convincing explanation for the blue shifts observed in the absorption and emission spectra in these materials.[29-32] The fact that Si and Ge have been well studied and are widely available should help make the integration of any new technologies based on nanoclusters of Si or Ge with those of the already existing technologies that much easier.[28]

Synthetic Methods. Methods that have been used to synthesize these clusters are of three different types. The most successful methods have used the gas-phase decomposition of silanes.[26,33,34] In this system, a series of higher silanes, disilanes and silylene polymeric species are formed at high temperatures (850-1050°C) in an enclosed vessel. There is little size control during the synthesis step of this high

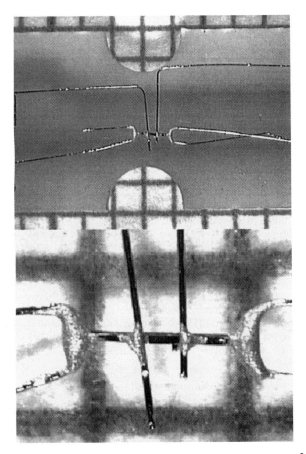

Figure 3. A single crystal of $Eu_{14}MnSb_{11}$ (0.006 x 0.006 x 1.000 mm^3) with 4 Pt leads attached with silver paint. The background grid is 1 mm^2.

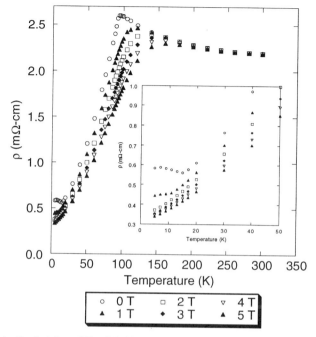

Figure 4. Resistivity of $Eu_{14}MnSb_{11}$ as a function of temperature and applied magnetic fields of 0T, 1T, 2T, 3T, 4T, and 5T. (Adapted from ref. 53)

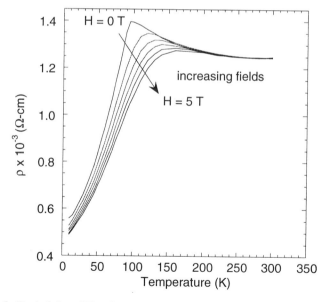

Figure 5. Resistivity of $Eu_{13}CaMnSb_{11}$ as a function of temperature and applied magnetic fields of 0T, 1T, 2T, 3T, 4T, and 5T.

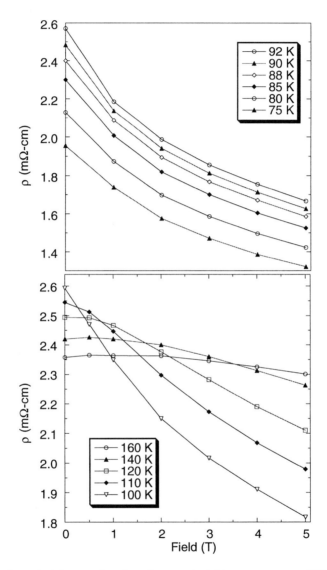

Figure 6. Resistivity as a function of magnetic field at temperature ≤ 92 K (top) and ≥ 92 K (bottom).

temperature process other than what can be achieved by changes in the initial reactant concentrations and the length of time over which agglomeration is allowed to take place. This results in a comparatively wide size distribution which can be greatly narrowed through use of size-selective precipitation.

A second method used by this group and others to synthesize silicon nanoparticles involves the ultrasonic dispersion of porous silicon in different solvents.[25,35,36] Changing the composition of the anodization solution, the HF concentration, the current density used during anodization, and the length of time of anodization can all be used to some degree to change the morphology and size of the crystallites. However, a wide range still results regardless of the set of parameters used. This is a relatively inexpensive method for producing silicon nanoparticles that luminesce, but the effort necessary to control size distribution and for quantitative characterization is prohibitive for large scale syntheses.

To date, the only reported solution phase synthesis for silicon nanocluster production, other than the one we are currently developing, is based on the reduction of $SiCl_4$ and $RSiCl_3$ by sodium metal in nonpolar organic solvents.[24] The synthesis is carried out at relatively high temperatures ($385°C$) and pressures (> 100 atm) with rapid stirring for 3 to 7 days. Analysis using transmission electron microscopy of the product from the reaction where R = hydrogen in $RSiCl_3$ reveals a size distribution of 5 to 3000 nm for silicon crystallites which are mostly hexagonally shaped. Most are single crystals but a few of the smaller crystallites are aggregates made up of two or three individual crystallites. The diameter of the various crystallites in this reaction only varies from 2 to 9 nm when R = octyl. The yield from this reaction is low at less than 10%.

As is the case with silicon, germanium nanoclusters have been produced using several different methods. None of these have become the synthetic method of choice but this may be due to the fact that the possibilities for nanocluster technology in germanium have not been explored as thoroughly as in silicon. Many of the methods to produce germanium nanoparticles are similar to those used in silicon nanocluster synthesis. One such synthesis uses nearly the same techniques as those used in the solution synthesis of silicon particles but substitutes the germanium analogs for the reactants.[37] The amount of sodium in the reaction solution has an effect on both the size of the particles produced as well as the shape. In addition to quantum dots, quantum wires and single crystalline platelets are produced. Unlike the silicon synthesis that produces mostly amorphous material as the product, almost all of the germanium product is crystalline. This may be due to the lower melting point of germanium which in turn requires a lower annealing temperature for crystallization to occur. If an excess of sodium is used, the yield of crystalline material is greatly reduced and only the quantum wires are produced.

Another method for producing germanium quantum dots involves their growth in SiO_2, GeO_2, or $SiNx$ matrices.[38-41] This technique can produce a wide variety of sizes and morphologies that depend on the particular parameters used. One major disadvantage of this technique is that the quantum dots formed are permanently imbedded within a solid matrix. This prevents any mechanical or chemical manipulation of these clusters.

Zintl precursor. A solution phase synthesis that produces silicon and germanium nanoclusters at much lower pressures and temperatures than those of previous solution phase methods has been achieved by use of the intermetallic Zintl salts NaSi, KSi, NaGe, and KGe as starting materials.[42-44] Zintl compounds have been explored as potential precursors in the synthesis of novel compounds.[45,46] Zintl compounds have also been shown to be useful for the synthesis of novel materials with potentially useful applications.[6,47,48] Although the ASi (A = Na, K) compounds have been known for some time,[49,50] there are few examples of their use as synthetic reagents.[51] The ASi and AGe (A = Na, K) compounds consist of covalently bonded

T_4^{4-} (T=Si, Ge) anionic clusters separated by 4 A^+ (A=Na, K) cations.[52] The anionic clusters are isostructural and isoelectronic with those of white phosphorus. Each of the group IV atoms in the anionic cluster possesses a charge of -1, for a total charge of -4 on each cluster. These clusters are suspended in an appropriate coordinating solvent and are reacted with the (formally) cationic Si^{4+} or Ge^{4+} of $SiCl_4$ or $GeCl_4$ respectively. The nanoclusters are Cl terminated and reaction with alkly lithium or alkyl Grignards produces a hydrophobic product that can be washed with water and extracted into a nonpolar organic solvent such as hexane.

Figure 7 shows the FTIR of the Ge nanoclusters, prepared in digylme, and terminated with octylMgBr. Several drops of the colloid in hexane were placed on a CsI plate. The solvent was allowed to evaporate and the FTIR is that of the dried colloid. The stretches are consistent with an alkyl group on the surface and assignments are provided on the figure.

Figure 8 shows the high resolution TEM micrograph of Ge nanoclusters on a amorphous carbon grid produced by the method described above. Most particles are between 2.0 and 5.0 nm in diameter. One nanocrystal is outlined and enlarged in Figure 8b. Lattice fringes can be clearly seen corresponding to the {111} planes ($\approx$ 3.27 Å) in germanium having the diamond structure. The selected area electron diffraction pattern corresponds to the {111} and {220} lattice planes of diamond structure germanium.

The development of research aimed at producing group IV nanoclusters has been presented. In this chapter, we have described the synthetic technique for producing Si and Ge nanoclusters. At the present time, we are working on optimizing yields and gaining control over the size distribution. More work needs to be done in order to more fully characterize these nanoclusters. Auger and EELS spectroscopy will provide more complete information on the surfaces. Photoluminescence and time resolved spectroscopy will be performed and compared with nanoclusters of Si and Ge produced by alternate routes. This method produces particles which lend themselves to the easy manipulation of their surfaces. The fact that this is a solution route at normal atmospheric pressure makes it an attractive method because of the ease with which they are made.

Acknowledgements. We thank D. J. Webb, H. W. H. Lee, and P. P. Power for useful discussions. We also thank the staff at the National Center for Electron Microscopy (NCEM) for useful discussion and assistance with the HRTEM. Work at the NCEM was performed under the auspices of the Director, Office of Energy Research, Office of Basic Energy Science, Materials Science Division, U. S. Department of Energy under contract # DE-AC-03-76SF00098. Funding from the National Science Foundation (DMR-9505565), the Campus Laboratory Collaboration Program of the University of California is gratefully acknowledged. S.M.K. acknowledges a Maria Goeppert-Mayer award from Argonne National Laboratory (ANL) (97-98). J.Y.C. and B.R.T. acknowledge Dissertation Year Fellowships from ANL and Lawrence Livermore National Laboratory (97-98), respectively.

References

1. *Chemistry, Structure and Bonding of Zintl Phases and Ions*; Kauzlarich, S. M., Ed.; VCH Publishers: New York, 1996.
2. Corbett, J. D. *Chem. Rev.* **1985**, *85*, 383.
3. Schäfer, H.; Eisenmann, B.; Müller, W. *Angew. Chem., Int. Ed. Engl.* **1973**, *12*, 694.
4. Schäfer, H.; Eisenmann, B. *Rev. Inorg. Chem.* **1981**, *3*, 29.
5. Schäfer, H. *J. Sol. State Chem.* **1985**, *57*, 97.

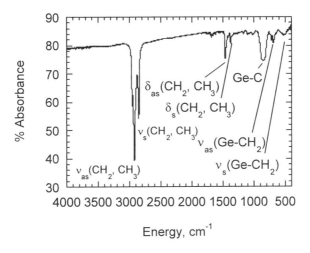

Figure 7. FTIR of Ge nanocrystals terminated with octylMgBr.

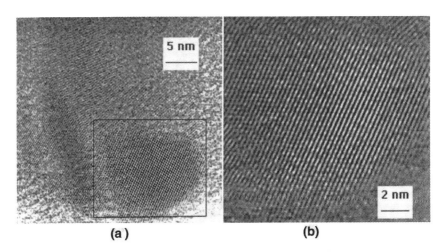

(a) (b)

Figure 8. (a) High resolution TEM micrograph of Ge nanocrystals on a amorphous carbon substrate. The accelerating voltage was 200 keV. One of the nanocrystals is outlined and the <111> face of Ge is imaged. (b) An enlargement of the outlined area. Both images have been rotationally filtered to remove background from the amorphous carbon support film.

26

6. O'Connor, C. J.; Jung, J.-S.; Zhang, J. H. In *Chemistry, Structure, and Bonding of Zintl Phases and Ions*; S. M. Kauzlarich, Eds.; VCH Publishers, Inc.: New York, 1996; pp 275.

7. Cordier, G.; Schäfer, H.; Stelter, M. *Z. anorg. allg. Chem.* **1984**, *519*, 183.

8. Kauzlarich, S. M.; Kuromoto, T. Y.; Olmstead, M. M. *J. Am. Chem. Soc.* **1989**, *111*, 8041.

9. Vidyasagar, K.; Hönle, W.; von Schnering, H. G. *Z. anorg. allg. Chem.* **1996**, *622*, 518.

10. Vaughey, J. T.; Corbett, J. D. *Chem. Mater.* **1996**, *8*, 671.

11. Carrillo-Cabrera, W.; Somer, M.; Peters, K.; von Schnering, H. G. *Chem. Ber.* **1996**, *129*, 1015.

12. Kauzlarich, S. M. In *Chemistry, Structure, and Bonding of Zintl Phases and Ions*; S. M. Kauzlarich, Eds.; VCH Publishers: New York, 1996; pp 245.

13. Chan, J. Y.; Wang, M. E.; Rehr, A.; Kauzlarich, S. M.; Webb, D. J. *Chem. Mater.* **1997**, *9*, 2131.

14. Gallup, R. F.; Fong, C. Y.; Kauzlarich, S. M. *Inorg. Chem.* **1992**, *31*, 115.

15. Young, D. M.; Torardi, C. C.; Olmstead, M. M.; Kauzlarich, S. M. *Chem. Mater.* **1995**, *7*, 93.

16. Kittel, C. *Solid State Physics* **1968**, *22*, 1.

17. Liu, L. *Sol. State Commun.* **1980**, *35*, 187.

18. Webb, D. J.; Kuromoto, T. Y.; Kauzlarich, S. M. *J. Magn. Magn. Mater.* **1991**, *98*, 71.

19. Kuromoto, T. Y.; Kauzlarich, S. M.; Webb, D. J. *Chem. Mater.* **1992**, *4*, 435.

20. Rehr, A.; Kuromoto, T. Y.; Kauzlarich, S. M.; Del Castillo, J.; Webb, D. J. *Chem. Mater.* **1994**, *6*, 93.

21. Parkin, S. S. P. *Annu. Rev. Mater. Sci.* **1995**, *25*, 357.

22. Ramirez, A. P. *J. Phys.: Condens. Matter* **1997**, *9*, 8171.

23. Canham, L. T. *Appl. Phys. Lett.* **1990**, *57*, 1046.

24. Heath, J. R. *Science* **1992**, *258*, 1131.

25. Heinrich, J. L.; Curtis, C. L.; Credo, G. M.; Kavanagh, K. L.; Sailor, M. J. *Science* **1992**, *255*, 66.

26. Littau, K. A.; Szajowshki, P. J.; Muller, A. J.; Kortan, A. R.; Brus, L. E. *J. Phys. Chem.* **1993**, *97*, 1224.

27. Lee, H. W. H.; Davis, J. E.; Olsen, M. L.; Kauzlarich, S. M.; Bley, R. A.; Risbud, S.; Duval, D. *Mat. Res. Soc. Symp. Proc.* **1994**, *351*, 129.

28. Brus, L. *J. Phys. Chem.* **1994**, *98*, 3575.

29. Brus, L. E.; Szajowski, P. F.; Wilson, W. L.; Harris, T. D.; Schuppler, S.; Citrin, P. H. *J. Am. Chem. Soc.* **1995**, *117*, 2915.

30. Wilson, W. L.; Szajowski, P. F.; Brus, L. E. *Science* **1993**, *262*, 1242.

31. Batson, P. E.; Heath, J. R. *Phys. Rev. Lett.* **1993**, *71*, 911.

32. Schuppler, S.; Friedman, S. L.; Marcus, M. A.; Adler, D. L.; Xie, Y.-H.; Ross, F. M.; Chabal, Y. J.; Harris, T. D.; Brus, L. E.; Brown, W. L.; Chaban, E. E.; Szajowshki, P. F.; Christman, S. B.; Citrin, P. H. *Phys. Rev. B* **1995**, *52*, 4910.

33. Fojtik, A.; Henglein, A. *Chem. Phys. Lett.* **1994**, *221*, 363.

34. Zhang, D.; Kolbas, R. M.; Mehta, P.; Singh, A. K.; Lichtenwalner, D. J.; Hsieh, K. Y.; Kingon, A. I. *Mat. Res. Soc. Symp. Proc.* **1992**, *256*, 35.

35. Bley, R. A.; Kauzlarich, S. M.; Lee, H. W. H.; Davis, J. E. *Mat. Res. Soc. Symp. Proc.* **1994**, *351*, 275.

36. Bley, R. A.; Kauzlarich, S. M.; Davis, J. E.; Lee, H. W. H. *Chem. Mater.* **1996**, *8*, 1881.

37. Heath, J. R.; LeGoues, F. K. *Chem. Phys. Lett.* **1993**, *208*, 263.

38. Fujii, M.; Hayashi, S.; Yamamoto, K. *Jpn. J. Appl. Phys.* **1991**, *30*, 687.

39. Kanemitsu, Y.; Uto, H.; Masumoto, Y.; Maeda, Y. *Appl. Phys. Lett.* **1992**, *61*, 2186.
40. Qu, X.; Chen, K. J.; Wang, M. X.; Li, Z. F.; Shi, W. H.; Feng, D. *Solid State Commu.* **1994**, *90*, 549.
41. Carpenter, J. P.; Lukehart, C. M.; Henderson, D. O.; Mu, R.; Jones, B. D.; Glosser, R.; Stock, S. R.; Wittig, J. E.; Zhu, J. G. *Chem. Mater.* **1996**, *8*, 1268.
42. Bley, R. A.; Kauzlarich, S. M. *J. Am. Chem. Soc.* **1996**, *118*, 12461.
43. Bley, R. A.; Kauzlarich, S. M. *A low Temperature Solution Phase Route for the Synthesis of Silicon Nanoclusters*; Kluwer Academic Publishers: 1996, pp 467.
44. Taylor, B. R.; Kauzlarich, S. M.; Lee, H. W. H.; Delgado, G. R. *Chem. Mater.* **1998**, *10*, 22-24.
45. Ahlrichs, R.; Fenske, D.; Fromm, K.; Krautscheid, H.; Krautscheid, U.; Treutler, O. *Chem. Eur. J.* **1996**, *2*, 238.
46. Charles, S.; Fettinger, J. C.; Eichhorn, B. W. *J. Am. Chem. Soc.* **1995**, *117*, 5303.
47. Haushalter, R. C.; Krause, L. J. *Thin Solid Films* **1983**, *102*, 2312.
48. Haushalter, R. C.; O'Conner, C. J.; Haushalter, J. P.; Umarji, A. M.; Shenoy, G. K. *Angew. Chem., Int. Ed. Engl.* **1984**, *23*, 169.
49. Zintl, E.; Goubeau, J.; Dullenkopf, W. *Z. Physikal. Chem.* **1931**, *154*, 1.
50. Zintl, E.; Kaiser, H. *Z. Anorg. Allg. Chem.* **1933**, *211*, 113.
51. Hey-Hawkins, E.; von Schnering, H. G. *Chem. Ber.* **1990**, *124*, 1167.
52. Schäfer, R.; Klemm, W. *Z. Anorg. Allg. Chem.* **1961**, *312*, 214.
53. Chan, J. Y.; Klavins, P.; Shelton, R. N.; Webb, D. J.; Kauzlarich, S. M. *Chem. Mater.* **1997**, *9*, 3132-3135.

Chapter 3

Metal Halide Framework Solids: Analogs of Aluminosilicates and Aluminophosphates

James D. Martin

Department of Chemistry, North Carolina State University,
Raleigh, NC 27695

A family of halozeotype materials have been constructed based on a structural analogy between $ZnCl_2$ and SiO_2. By substituting Cu^I for certain Zn^{II} cations, frameworks can be created with the general formula $[A]_n[Cu_nZn_{m-n}Cl_{2m}]$, analogous to the extensive family of aluminosilicate zeolites, $[A]_n[Al_nSi_{m-n}O_{2m}]$. Similar construction principles can be applied to the synthesis of copper aluminum halides of general formula $Cu_mAl_mCl_{4m}$, analogous to the family of aluminophosphates, $Al_mP_mO_{4m}$. Both known and novel zeolite-type structures have been characterized. These metal halide frameworks are much less robust, and more chemically reactive than the corresponding refractory metal oxides. In addition, the structural relationship between the halide and oxide frameworks is consistent with a remarkable framework flexibility which is implicated in the adsorption of small molecules.

Silicates, aluminosilicates, and aluminophosphates are materials which have diverse utility in ion exchange, molecular sieve, sensor and catalysis applications. This is in large part due to the ability design new porous crystalline frameworks in which the crystal structure is tailored to specific functions. In these silicate-type frameworks, tetrahedral $TO_{4/2}$ building blocks (T = tetrahedral atom), are readily assembled around molecular and/or ionic templates, linked by two coordinate oxygen bridges. More recently, several groups have expanded these principles of framework construction to a variety of non-oxide crystal systems (1). Our particular interest is to utilize the principles of silicate-type framework synthesis for the construction of materials in which size and shape selective frameworks are built out of reactive building blocks such as metal halides. While this work describes the design of metal-halide materials related to $ZnCl_2$, the same principles can be applied to any metal dianion, TX_2, in which the metal has a preference for tetrahedral coordination, whereby it provides a silicate-type parentage to a family of framework materials.

The relationship between $ZnCl_2$ and SiO_2 was previously observed in the study of their respective glasses (2,3). While there are limits to the analogy, $ZnCl_2$ was to some extent considered a low temperature model of silicate glasses. Nevertheless, little effort has been made to exploit the structural analogy between these materials. We have found that by substitution of certain of the Zn^{II} sites of $ZnCl_2$ by Cu^I, anionic frameworks, $[Cu_nZn_{m-n}Cl_{2m}]^{n-}$, can be created around cationic templates analogous to the synthesis of aluminosilicates $[Al_nSi_{m-n}O_{2m}]^{n-}$. Similarly, syntheses utilizing Cu^I and Al^{III} tetrahedral cations fulfills the same charge requirements as two Zn^{II} cations thus affording materials with the general formula $Cu_mAl_mCl_{4m}$ which are analogs to aluminophosphates, $Al_mP_mO_{4m}$.

Synthesis

Because the metal halide building units of these novel framework materials generally have much lower melting points than the corresponding oxides, relatively mild synthetic conditions afford a rich chemistry. In addition, binary or ternary mixtures normally melt at temperatures significantly below that of the individual reagents and frequently well below the decomposition temperature of many organic molecules which are useful as framework templates. All of the materials described here can be prepared by a cooling of the melt (~250°C) of stoichiometric mixtures of their respective components. We have also found that excellent single crystal growth of these materials can generally be achieved by solvothermal reactions using superheated benzene as solvent (~160°C) in thick walled fused silica ampoules. To safely utilize such reaction ampoules under the generated high pressures, they should never be filled to more than 60 to 80% of the ampoules volume. In addition, because the metal halides are air and moisture sensitive, all manipulations were carried out under and inert atmosphere.

The Silicate-like Parentage of $ZnCl_2$

To begin this discussion, it is instructive to examine the structural relationships between the numerous polymorphs of $ZnCl_2$ and SiO_2. Of the four crystalline forms of $ZnCl_2$ that have been characterized to date (4,5), α-$ZnCl_2$ is the only direct silicate structural analog; related to high cristobalite (6). The structure of orthorhombic-$ZnCl_2$ is analogous to a predicted polymorph of SiO_2 (7) and is also observed for several silicate type ternaries, ABX_2, and quaternaries, $A_2BB'X_4$, that are derived from a filling of this framework (6). Both α-$ZnCl_2$ and high cristobalite crystallize in the tetragonal space group $I\bar{4}2d$. However, the most stark contrast between these structures, apparent in Figure 1, is the arrangement of the anions into ordered layers in $ZnCl_2$ as opposed to the puckered layers in SiO_2. This puckering of the anion layers in SiO_2 is a function of the expanded Si-O-Si angle of 145° due to the short Si-O distance. In $ZnCl_2$ the metal anion distance is longer, which allows for a more contracted Zn-Cl-Zn angle, 109°, before experiencing cation-cation (T-T) repulsions. As will be described later, the possibility of framework expansion resulting from an expansion of this T-X-T angle in the halide structures creates structural flexibility which is useful for the adsorption of small molecules. Interestingly, the contraction

a.

b.

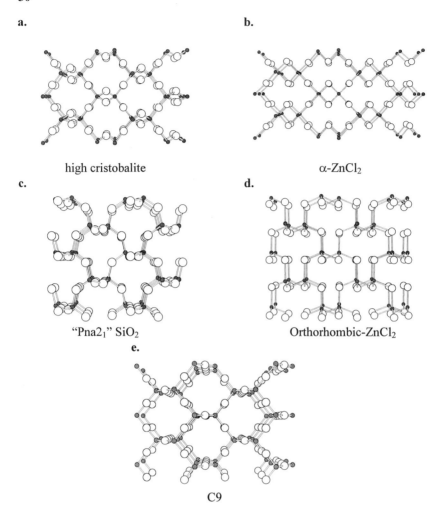

high cristobalite

α-ZnCl₂

c.

d.

"Pna2₁" SiO₂

Orthorhombic-ZnCl₂

e.

C9

Figure 1: Ball and stick drawings indicating the similarity between SiO₂ (a. and c.) and ZnCl₂ (b. and d.) polymorphs. The predicted structure of Pna2₁ SiO₂ is equivalent to the real structure of ZnSiO₄. (e) Ball and stick representation of the C9 diamondoid structure.

of the T-X-T angle to 109° in ZnCl₂ brings the anions in to a closest packed arrangement in which the metal cations fill 1/4 of the tetrahedral holes in an ordered 1 X 1 zigzag chain pattern as shown in Figure 2. This layer represents a (112) slice of α-ZnCl₂ or a (001) slice of orthorhombic-ZnCl₂. By stacking these layers in a cubic closest packed fashion, ...ABCA... the α-ZnCl₂ structure is obtained whereas the orthorhombic-ZnCl₂ polymorph is obtained by hexagonal closest packing,

...ABA.... Further it is noted that both of these structures are derivatives of the same diamondoid, C9 structure-type, which can be obtained starting from either α-ZnCl$_2$ or orthorhombic-ZnCl$_2$ by opening the T-X-T angle to 180° as shown in Figure 1e.

a.   **b.**

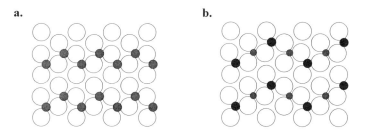

Figure 2. Representation of the cation ordering in the closest packed layers of (a) ZnCl$_2$ (a 112 section of α-ZnCl$_2$ and a (010) section of orthorhombic-ZnCl$_2$) and (b) CuAlCl$_4$ (a 112 section of α-CuAlCl$_4$ and a (001) section of β-CuAlCl$_4$). In b. the smaller lighter shaded spheres are Al and the larger, darker spheres represent Cu.

Cupric Haloaluminates: Analogs of Aluminophosphates

A direct analog of the α-ZnCl$_2$ structure type is found in the structure of CuGaI$_4$ (8) similar to the structural relationship between SiO$_2$ and AlPO$_4$. Here the CuI$_{4/2}$ tetrahedra, Cu-I = 2.62Å , and the GaI$_{4/2}$ tetrahedra, Ga-I = 2.56Å, are virtually of identical size and thus the +1 and +3 cations directly substitute for two Zn^{II} cations. However, in CuAlCl$_4$ (9,10), the CuCl$_{4/2}$ tetrahedra are significantly larger, Cu-Cl = 2.36Å, than the AlCl$_{4/2}$ tetrahedra, Al-Cl = 2.14Å. This difference in the size of the tetrahedral building blocks has a more profound effect on the cation arrangement in the closest packed halides than is observed for the oxide analogs (for example AlPO$_4$), such that in CuAlCl$_4$ a 2 x 2 zigzag chain arrangement of the cations in the tetrahedral holes of the closest packed anion layers is observed (shown in Figure 2). Stacking of these closest packed layers in a cubic closest packed fashion results in the formation of α-CuAlCl$_4$ whereas a hexagonal closest packed stacking of the layers results in β-CuAlCl$_4$, as shown in Figure 3. The α-CuAlCl$_4$ crystallizes in the tetragonal space group $P\bar{4}2c$, **a** = **b** = 5.4409(1)Å and **c** = 10.1126(3)Å, and β-CuAlCl$_4$ crystallizes in the orthorhombic space group $Pna2_1$, **a** = 12.8388(5)Å, **b** = 6.6455(3)Å, and **c** = 6.1264(3)Å. The β-phase of CuAlCl$_4$ has been shown to be metastable with respect to the α-phase and can only be synthesized by rapid quenching from a melt (9). The β-phase is stable up to about 100°C at which point it reverts to the more thermodynamically stable α-phase.

The ordering of the cations into zigzag chains, occupying only 1/4 of the tetrahedral holes, creates a lattice with notable "van der Waals" channels throughout the framework. These channels are physically too small for penetration by small molecules; the cross channel Cl-Cl contacts being 0.2 to 0.4Å greater than the sum of the van der Waals radii of Cl. Nevertheless, we have shown by gravimetric

a.

b.

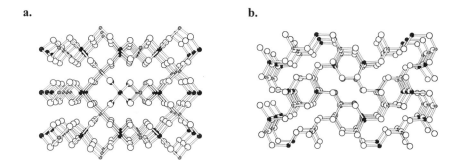

Figure 3. Ball and stick representations of (a) α-CuAlCl₄ looking down **a**, and (b) β-CuAlCl₄ looking down **c**. The non-shaded spheres are Cl, the smallest lighter shaded spheres are Al and the larger, darker spheres represent Cu.

analysis, and spectroscopically, that α-CuAlCl₄ can reversibly adsorb 0.5 to 1.0 molar equivalents of small molecule gasses such as C_2H_4, CO and NO (9,11). This gas sorption also may be related to the reversible benzene adduct formation of $(C_6H_6)CuAlCl_4$ shown in Figure 4a (12, 13). Upon gas desorption the metastable β-phase is completely converted to the α-phase. In the case of the ethylene adsorption, X-ray powder diffraction confirms that a tetragonal phase with lattice constants of

a.

b.

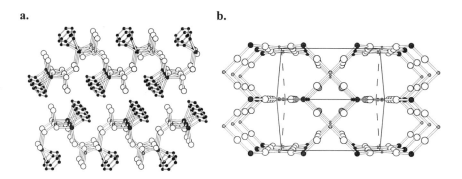

Figure 4. (a) Ball and stick representation looking parallel to the layers in $(C_6H_6)CuAlCl_4$. (b)Ball and stick representation of the proposed expanded tetragonal structure, looking down the 110 direction, observed for the ethylene adduct of α-CuAlCl₄. The position of the ethylene is likely coordinated to the Al (small shaded spheres). The $CuCl_{4/2}$ unit remains tetrahedrally coordinated.

$\mathbf{a} = \mathbf{b} = 9.0$ and $\mathbf{c} = 11.9$Å is formed (11). This powder pattern is consistent with the model shown in Figure 4b in which the T-X-T angles in the α-phase are expanded from the observed 110° to near 180° while maintaining framework integrity. We have not yet determined the position of the adsorbed olefin within this model, however, the X-ray powder diffraction can be indexed to a $\sqrt{2}$ x $\sqrt{2}$ superstructure of the [NMe$_4$][CuPt(CN)$_4$]-type structure in which the olefins may well be coordinated to the Al giving it an octahedral coordination geometry. Such olefin binding to aluminum has precedence in the molecular literature (15). However, other preliminary work indicates that other small molecules can enter this framework upon adsorption and coordinate to the Cu after cleavage of a Cu-Cl bond (11, 13). Nevertheless, while various structures are formed by the small molecule adsorption into the CuAlCl$_4$ framework, in all cases, the reversible sorption process appears to be dictated in part by the flexibility of the metal halide frameworks.

Copper Zinc Halides: Analogs of Aluminosilicates

In the materials described above, the "micro-porosity" was completely dependent on a remarkable flexibility of the frameworks, as opposed to the engineering of framework porosity by templated growth. By contrast, construction by substituting Cu$^+$ for certain of the Zn^{2+} sites in ZnCl$_2$ lattices results in the anionic frameworks [Cu$_n$Zn$_{m-n}$Cl$_{2m}$]$^{n-}$, analogs of framework aluminosilicates, in which extra-framework cations can be exploited as templates for framework design. The more mild conditions required for synthesis and crystal growth of these metal halide materials may in fact allow for the geometry of the template to exhibit a more direct influence on the framework structure than is often observed for the more refractory oxides.

CZX-1 The Sodalite Structure. The trimethylammonium cation serves to template the formation of β-cages resulting in the ubiquitous sodalite structure, [HNMe$_3$][CuZn$_5$Cl$_{12}$], CZX-1 shown in Figure 5 (16). (CZX is defined as the common name for this family of **C**opper **Z**inc Halides.) Crystallizing in the cubic space group $I\bar{4}3m$, with $\mathbf{a} = 10.5887(3)$Å, the trimethylammonium template is disordered about four equivalent positions within the cage. Remarkably, CZX-1 exhibits the most contracted T-X-T angle (110.04(3)°) of any known sodalite structure. This is in marked contrast to oxide based sodalites in which the T-O-T angles are known to range from 124-160° (17). The minimum angle is limited by the short T-O bond distances which, for small T-O-T angles, force short T-T contacts. The structural comparison between the oxide and halide based sodalite structures is parallel to that described above for ZnCl$_2$ and SiO$_2$, and is similarly evident in the respective anion arrangements shown in Figure 5. With the nearly ideal tetrahedral angle at Cl, the halide based sodalite structure clearly shows a cubic closest packing arrangement of the framework anions. And the role of the template is to fill 4/16 of the closest packed anion sites such that the β-cages could be formed (Figure 5c). We have not yet demonstrated framework flexibility of this cage, nevertheless, this most contracted metal-halide β-cage (centroid to Cl distance is 4.41Å) is of a similar size

34

to the most expanded oxide cages with T X-T = 156° and centroid - O = 4.41Å. Larger cations or other solvent molecules should readily be accommodated by increasing the T-Cl-T angle.

a.   b.   c.

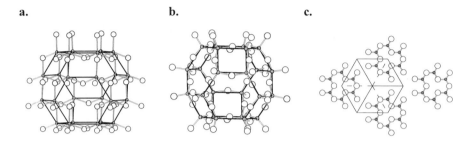

Figure 5. Ball and stick representations of (a) CZX-1 and (b) $Al_3Si_3O_{12}^{3-}$ emphasizing the distortion of the anion layers in the oxide material. And (c) a (111) section of CZX-1 emphasizing the 4/16 vacancies in the ccp anion sublattice.

CZX-2 and CZX-3 A Novel Microporous Framework. Syntheses utilizing the more rod like diethyl- or dimethylammonium cations results in the templated growth of a novel zeotypic framework with a three dimensional channel structure shown in Figure 6. CZX-2, $[H_2NEt_2][CuZn_5Cl_{12}]$, and CZX-3, $[H_2NMe_2]_n[Cu_nZn_{6-n}Cl_{12}]$ (n = 1 or 2) crystallize in the orthorhombic space group $I2_12_12_1$ with **a** = 9.6848(5) Å, **b** = 9.5473(4) Å and **c** = 14.0003(9)Å and **a** = 9.5677(16) Å, **b** = 9.4554(12) Å and **c** = 13.6435(16)Å, for CZX-2 and CZX-3 (n = 2) respectively (16). Consistent with our aluminosilicate analogy, CZX-2 and CZX-3 are constructed from corner sharing tetrahedral primary building units. Like the halozeotypes described above, these materials also exhibit T-X-T angles of 110°; well contracted below that observed for any aluminosilicates. Nevertheless, the connectivity of this framework can not be described from a closest packing of anions.

a.   b.   c.

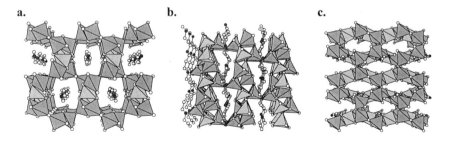

Figure 6. Polyhedral representation of the framework of CZX-2 as viewed (a) along **b**, (b) down the 111 vector, and (c) along **c**. The diethylammonium cations are shown within the framework. However, a 50% occupancy by H_2NEt_2 is crystallographically required.

The most open direction of this framework consists of channels running along the 2_1 axes parallel to **b**. The smallest ring to circumscribe this channel consists of eleven tetrahedra. As shown in Figure 6 (for CZX-2), the templating dialkylammonium cations reside in the center of these channels, and are hydrogen bonded to the chloride channel interior. In the stuffed n = 2 CZX-3, the cations are oriented in a slightly different fashion within the same channels in order to maximize hydrogen bonding. The shortest cross channel contacts are 6.34Å for CZX-2 which corresponds to a free diameter of 2.9Å given the van der Waals radii of Cl = 1.7Å. Four 8-ring windows provide entry into the primary 11-ring channels forming smaller channels along the body diagonal. The smallest cross channel contacts of this 8-ring channel are 4.32Å in CZX-2. The largest remaining void in the fully occupied CZX-3 structure is centered directly in the 8-ring window, and also corresponds to the channel parallel to the **c** axis (shown in the Figure 6) with the shortest void centroid-to-halide distance of 2.69Å. This void space, in addition to the void created by template vacancy in CZX-3 (n = 1) are of appropriate size to adsorb small molecules such as methanol. While this solvent adsorption also results in colloid formation, we have shown that CZX-3 (n = 1) adsorbs 1.8 moles per formula unit more methanol than does its stuffed n = 2 counterpart.

CZX-4: an Analog to Oxides or Sulfides? Use of the smaller templating cations, H_3NMe^+ or Rb^+, yields the framework $[A][Cu_2Zn_2Cl_7]$, CZX-4, which begins to deviate from the silicate analogy (18). This framework also is constructed out of corner sharing tetrahedral building units, however one of the chloride anions is three coordinated accounting for the lower anion/T-atom ratio. Like CZX-1, the template effectively fills the space of certain anion sites in a closest packed lattice. As shown in Figure 7, 1/8 of the anion sites in a closest packed layer are occupied by the extra-framework cation. These layers are then stacked in a hexagonal closest packed fashion such that the template resides in a cage of twelve chloride anions (See Figure 8). The structure of the frameworks is virtually identical for the methylammonium (CZX-4) and rubidium (Rb-CZX-4) templated materials. However, the crystal

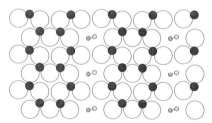

Figure 7. Representation of the closest packed layer in CZX-4 with templating cations in 1/8 of the anion sites.

symmetry is monoclinic *Pn*, for CZX-4, **a** = 6.3098(8)Å, **b** = 6.339(8)Å and **c** = 15.569(2)Å, but orthorhombic *Pmn2₁*, **a** = 15.238(1)Å, **b** = 6.3972(4)Å and **c** = 6.1305(11)Å for Rb-CZX-4 because of the respective rod-like and spherical symmetry of the templates.

a.   **b.**

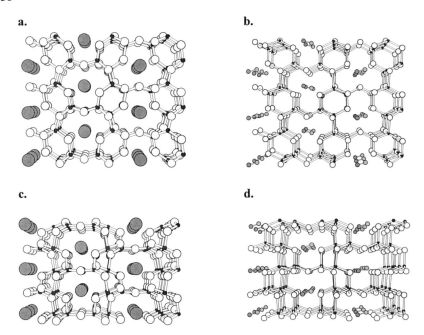

c.   **d.**

Figure 8. Ball and Stick drawings viewed perpendicular to (a. and b.) and parallel to (c. and d.) the "closest packed" anion planes of SrB_4O_7 (a. and c.) and $[H_3NMe]Cu_2Zn_2Cl_4$, CZX-4 (b. and d.), respectively.

While no silicates, or aluminosilicates are known to adopt this structure, an identical framework is observed for $BaAl_4S_7$ and $BaGa_4S_7$ (19). Like CZX-4, the anion sublattice in both of these sulfides is nearly ideally hexagonal closest packed, with the extra-framework cation residing in 1/8 of the anion sites. As initially described for $ZnCl_2$, above, the closest packed geometry is accessible when the T-X bond distance is long enough to allow the T-X-T angle to achieve approximately 109° without creating T-T repulsions. This is in fact observed for these sulfides with Al-S-Al = 105° to 109° as compared to 110° to 114° in CZX-4 and 105° to 110° in Rb-CZX-4. By contrast, oxide analogs to this structure type only exist for the smaller boron T-atoms, for example, SrB_4O_7 (20), EuB_4O_7 (21) and PbB_4O_7 (22). Yet even here, the short B-O bond distance require a significant expansion of the T-X-T angle to a range of 115° to 140°. This T-X-T angle expansion, again results in significant distortion to the framework, as seen in Figure 8, but also offers the possibility of significant framework flexibility.

Conclusions

The principles of non-oxide silicate-type framework construction are demonstrated based on an analogy between $ZnCl_2$ and SiO_2. Extension of this analogy was demonstrated by the preparation of $Cu_mAl_mCl_{4m}$, analogs of aluminophosphates, and

$[A]_n[Cu_nZn_{m-n}Cl_{2m}]$ analogs of aluminosilicates. In the copper zinc chlorides, the extra-framework cations exhibit a clear templating influence giving rise to both known and novel zeotype framework structures. Structurally, the most notable difference between the halide and oxide framework structures is the variation in the T-X-T angles. It is possible for this angle to be approximately 109° in the metal halide structures as a result of the longer M-Cl distances. This allows for the construction of open frameworks based on a closest packing of spheres model. Similar geometric parameters were also observed for $BaAl_4S_7$, suggesting a closer analogy between halides and sulfides than oxides. In oxides, the shorter M-O distances require an expansion of the T-X-T angles resulting in more distorted frameworks. In addition, while these halide frameworks are nowhere near as physically robust as the refractory metal oxides, a remarkable framework flexibility was demonstrated for the metal halides which allows for the adsorption of small molecules.

Acknowledgments.
This work was supported by the National Science Foundation, CAREER award, (DMR-9501370). J. D. Martin is a Cottrell Scholar of the Research Corporation.

References

1. *Reviewed by* Bowes, C. L; Ozin, G. A. Adv. Mater., 1996, 8, 13.
2. Angell, C. A.; Wong, J. J. Chem. Phys., 1970, 53, 2053.
3. Wong, J.; Lytle, F. W. J. Non-Cryst. Solids, 1980, 273.
4. Brehler, B. Z. Krist., 1961, 115, 373.
5. Yakel, H. L.; Brynestad, J. Inorg. Chem., 1978, 17, 3294.
6. O'Keeffe, M.; Hyde, B. G. Acta. Cryst., 1976, B32, 2923.
7. Boisen, M. B.; Gibbs, G. V.; Bukowinski, M. S. T. Phys. Chem. Min., 1994, 21, 269.
8. Burns, R.; Zajonc, A.; Meyer, G. Z. Krist., 1995, 210, 62.
9. Martin, J. D.; Leafblad, B. R.; Sullivan, R. M.; Boyle, P. D. Inorg. Chem., 1998, 37, 1341.
10. Hildebrandt, K.; Jones, P. G.; Schwarzmann, E.; Sheldrick, G. M. Z. Naturforsch., 1982, 37b, 1129.
11. Martin, J. D.; Sullivan, R. M. North Carolina State University, unpublished data.
12. Turner, R. W.; Amma, E. L. J. Am.Chem. Soc., 1966, 88, 1877.
13. Martin, J. D.; Leafblad, B. R.; Sullivan, R. M. North Carolina State University, unpublished data.
14. Gable, R.; Hoskins, B. F.; Robson, R. J. Chem. Soc., Chem. Commun., 1990, 762.
15. Fisher, J. D.; Wei, M.-Y.; Willett, R.; Shapiro, P. J. Organometallics, 1994, 13, 3324.
16. Martin, J. D.; Greenwood, K. B. Angew. Chem. Int. Ed. Eng., 1997, 36, 2702.

38

17. Depmeier, W. Acta. Cryst., 1984, B40, 185.
18. Greenwood, K. B.; Euliss, L. E.; Martin, J. D. Inorg. Chem., submitted.
19. Eisenmann, B.; Jakowski, M.; Schäfer, H. Rev. Chimie Min., 1983, 20, 329.
20. Krogh-Moe, J. Acta. Chem. Scand., 1964, 18, 2055.
21. Machida, K. I.; Adachi, G. Y.; Shioka, J. Acta Cryst., 1980, B36, 2008.
22. Corker, D. L.; Glazer, A. M. Acta. Cryst., 1996, B52, 260

Chapter 4

Chiral Chelate Complexes as Templates for Inorganic Materials

Angus P. Wilkinson

School of Chemistry and Biochemistry, Georgia Institute of Technology, Atlanta, GA 30332–0400

The use of simple Co(III) chiral chelate complexes as templates for the synthesis of aluminophosphate and gallophosphate materials is discussed. Chirality can be transferred from templates of this type to a growing inorganic framework. Co(III) cage complexes of the sarcophagine type and simple Ir(III) chelate complexes are shown to have much better hydrothermal stability than simple Co(III) complexes. The importance of electrostatics, space filling, and hydrogen bonding interactions in determining a material's structure and properties is outlined.

The demand for optically pure organic compounds is large, as such compounds are of great value in the production of pharmaceuticals and insecticides. The pharmaceutical industry, in particular, faces regulatory pressure to move away from formulations containing racemic mixes to ones containing the active agent in optically pure form. In 1996 sales of optically pure drugs in dosage form were in excess of $72 billion.*(1)* The synthesis of optically pure compounds to meet these needs clearly requires the use of chiral reagents, catalysts or enantioselective separations technology.

The development of chiral zeolitic materials in optically pure form would open up exciting new opportunities in enantioselective catalysis and separations science. Although, it is relatively easy to postulate plausible chiral zeolite frameworks, there has been very little success in preparing and controlling the chirality of these materials.*(2)* Newsam *et al.(3,4)* and Higgins *et al.(5)* have characterized a material, zeolite β, that is an intimate intergrowth of several polymorphs, one of which (polymorph A) is chiral. Attempts to prepare zeolite β samples containing large amounts of one enantiomer of the chiral polymorph have only succeeded in producing samples with a small enantiomeric excess (ee) in the intergrowth.*(6)* It has been demonstrated that this small ee within the zeolite catalyst sample is capable of producing diol products, from the hydrolysis of *trans*-stilbene oxide, that are enriched in one enantiomer.*(6)* The microporous titanosilicate ETS-10 also has a chiral polymorph.*(7)* Unfortunately, like zeolite β, it usually contains a large number of stacking faults and the chiral polymorph has never been prepared in pure form. A number of other open framework chiral materials have been prepared. For example, Goosecreekite,*(8)* NaZnPO$_4$·H$_2$O,*(9)* and a series of cobalt phosphates*(10)* have been reported, but no attempt has been made to control the chirality of these materials.

There is presently little experimental evidence for enantioselective separations or catalysis using zeolites with intrinsically chiral frameworks. However, Hutchings and Willock*(11)* have reported simulations examining the behavior of 2-butanol and the enantiomeric diols derived from the hydrolysis of *trans* stilbene oxide in purely siliceous polymorph A of zeolite β. Their work suggests that, for the diols, there is a significant difference in the binding energies of the two enantiomers and that for butan-2-ol the diffusion coefficients of the two enantiomers within the chiral zeolite are different. Additionally, there have been reports of enantioselective reactions using achiral zeolites containing chiral modifiers. For example, Faujasite type zeolites modified with chiral dithiane oxides have been experimentally shown to be effective for the enantioselective dehydration of 2-butanol,*(12-18)* and a variety of chiral natural products, including ephedrine, have been used in the chiral modification of zeolites that were used as hosts for enantioselective photochemical processes.*(19,20)*

Although the inclusion of chiral metal complexes in the cavity system of a zeolite can lead to supported catalysts with good enantioselectivity,*(21,22)* the production of highly enantioselective catalysts based upon intrinsically chiral zeolite frameworks is likely to difficult since framework structure can not be readily tailored to a particular reaction. However, the requirements for an industrially useful enantioselective separation using an intrinsically chiral zeolite are not as demanding of the material. The high pore volume of zeolites combined with modest enantioselective binding or diffusion could lead to viable large scale separations processes.

Syntheses of zeolitic aluminosilicates and metal phosphates often employ a molecular template or structure directing agent to produce the framework of interest. The preparation of new framework topologies is closely linked to the chemist's ability to devise new templates or structure directing agents. Typically, amines or alkylammonium salts are used in this role.*(2,23-25)* However, the use of chiral amines and alkylammonium ions has not been a very successful approach to the synthesis of chiral zeolitic materials. The synthetic opportunities offered by potential template species such as organometallic cations and coordination complexes have only just begun to be explored. As coordination complexes can be prepared with a wide variety of sizes, charges and shapes, and as chirality is easily achieved, we believe that they offer many interesting possibilities for preparing new structures. The rigidity of metal chelate complexes may be advantageous for the transfer of chirality from the template to the growing framework.

Inorganic Structure Directing Agents in Framework Syntheses

The use of metal complexes and organometallic species as structure directing or templating agents has received little attention. This is in part due to the difficulty of finding species that are stable under the conditions needed for framework synthesis. There have been reports describing the use of metallophthalocyanines,(26-29) $Ru(bipy)_3^{2+}$,(30) $Co(en)_3^{3+}$,(31,32), $Co(NH_3)_6^{3+}$,(33) $CoCp_2^+$,(34-36) $CoCp*_2^+$ (37,38) and a range of other chelate complexes(39) as additives to molecular sieve synthesis gels. The recent preparation of the first zeolite to contain 14 T-atom pores(38) using decamethylcobaltacenium as a structure directing agent is likely to arouse considerable further interest in the use of inorganic templates.

Simple Chiral Chelate Complexes as Templates. While a large number of chiral chelate complexes have been prepared, very few of them have the hydrothermal stability that is needed. Species containing a metal ion with a d^6 electron configuration are attractive as they typically display very slow ligand exchange kinetics and, consequently, are resistant to decomposition and racemization.

Our starting point was a series of Co(III) complexes with N-donor ligands. There is a large body of literature on this class of complex and many of them can be readily prepared and resolved. Initial experiments made use of en- 1,2-diaminoethane, tn - 1,3-diaminopropane and dien - diethylenetriamine as ligands. Complexes with these ligands show reasonable kinetic resistance to decomposition under acidic conditions and were used to prepare the aluminophosphate and gallophosphate materials listed in Table I *via* hydrothermal synthesis procedures. The three geometrical isomers of $Co(dien)_2^{3+}$ (designated *trans*, *u-cis*, and *s-cis*)(40,41) were investigated so that the influence of subtle changes in the template could be examined.

| *trans* | *s-cis* | *u-cis* |

Figure 1. The geometrical isomers of $Co(dien)_2^{3+}$

Of the complexes represented in Table I, only $Co(en)_3^{3+}$ was used as a single enantiomer. It exhibits good resistance to racemization and is easily resolved. $Co(tn)_3^{3+}$, and *trans*-$Co(dien)_2^{3+}$(42) are not as stable towards racemization, *u-cis*-$Co(dien)_2^{3+}$ is tedious to prepare in optically pure form and *s-cis*-$Co(dien)_2^{3+}$ is achiral.

Our syntheses of the metal phosphates listed in Table I demonstrated that the hydrothermal stability of the template species is dependent upon the nature of the ligands bound to the metal center. This observation was reinforced by subsequent experiments using complexes containing the ligands chxn and pn (chxn - *trans*-1,2-diaminocyclohexane, pn - 1,2-diaminopropane) in AlPO syntheses. $Co(pn)_3^{3+}$ is far more susceptible to decomposition than $Co(en)_3^{3+}$ and the stability of $Co(chxn)_3^{3+}$ is sufficiently bad that we were unable to prepare aluminophosphate single crystals containing this template.

Table I. Materials Prepared using Co(III) Complexes with en, tn and dien as Ligands

Added template	Product	ID	Notes	Ref
d,l-Co(en)$_3$$^{3+}$	[d,l-Co(en)$_3$][Al$_3$P$_4$O$_{16}$]xH$_2$O	1	Layered - type A	(43)
d,l-Co(en)$_3$$^{3+}$	[d,l-Co(en)$_3$][Ga$_3$P$_4$O$_{16}$]xH$_2$O	2	Layered - type A	(44)
d-Co(en)$_3$$^{3+}$	[d-Co(en)$_3$][Al$_3$P$_4$O$_{16}$]xH$_2$O	3	Layered - type B	(45)
d-Co(en)$_3$$^{3+}$	[d,l-Co(en)$_3$][H$_6$Al$_2$P$_6$O$_{24}$]$^{\#}$	4	AlPO cluster	(46)
d-Co(en)$_3$$^{3+}$	[d-Co(en)$_3$][H$_3$Ga$_2$P$_4$O$_{16}$]xH$_2$O	5	3D GaPO framework	(47)
d,l-Co(tn)$_3$$^{3+}$	[d-Co(tn)$_3$][Al$_3$P$_4$O$_{16}$]xH$_2$O*	6	Layered - type C	(48)
d,l-$trans$-Co(dien)$_2$$^{3+}$	[d-$trans$-Co(dien)$_3$][Al$_3$P$_4$O$_{16}$]xH$_2$O*	7	Layered - type B	(49)
s-cis-Co(dien)$_2$$^{3+}$	[s-cis-Co(dien)$_3$][Al$_3$P$_4$O$_{16}$]xH$_2$O	8	Layered - type B	(50)
u-cis-Co(dien)$_2$$^{3+}$	[u-cis-Co(dien)$_2$]$^{3+}$ AlPO	9	structure unknown	(50)

* - both enantiomers of the chiral solid are present in the product
# - template racemization occurred during synthesis

Template Species with Enhanced Hydrothermal Stability As all of the simple Co(III) chelate complexes showed poor base stability at elevated temperatures, and the ones with bulky ligands displayed poor stability even in acidic media, we examined alternative template species. Two strategies were pursued: i) the use of Ir(III) instead of Co(III), because of the significantly slower ligand exchange kinetics displayed by this metal ion, and ii) the use of ligands that encapsulate the metal center.

There is only a small body of literature on the synthesis and resolution of Ir(III) *tris*-chelate complexes with N-donor ligands. As Galsbøl had reported procedures for the preparation of Ir(en)$_3$$^{3+}$(51) and Ir(chxn)$_3$$^{3+}$,(52,53) we chose to start our exploration with these species. Under AlPO synthesis conditions these templates showed considerably better stability than the corresponding Co(III) complexes. We were able to perform syntheses at temperatures > 50°C above those that could be employed with the corresponding cobalt complexes. So far, we have produced two AlPOs with these templates, [d,l-Ir(en)$_3$][Al$_3$P$_4$O$_{16}$]·xH$_2$O (**10**) and [Ir(chxn)$_3$][Al$_2$P$_3$O$_{12}$]·4H$_2$O (**11**).

We chose to explore cage complexes of the sarcophagine and sepulchrate type because a wide variety of these complexes can be prepared by capping transition metal *tris*-chelates complexes, and, in many cases, they have good hydrothermal stability. For example, it has been reported that Co(sep)$^{3+}$ (sep - sepulchrate) does not decompose or racemize when boiled for prolonged periods in 12 M HCl.(54) Only about 40% of a sample of Co(diAMchar)$^{3+}$ decomposed after 3 days at 200 °C in 5M NaOH, and after 24 hours in 12M HCl at 275 °C there was negligible decomposition of a Co^{II}(diAMcharH$_2$)$^{4+}$ sample (diAMchar - capped *tris-trans*-1,2 diaminocyclohexane complex).(55)

We focused our initial work on dinitro and diamino sarcophagines (see Figure 2) rather than sepulchrates as cobalt (III) dinitrosarcophagine is trivially prepared from Co(en)$_3$$^{3+}$, and the dinitro compound is easily reduced to give the diamino complex.(56) Co(diNOsar)$^{3+}$ and Co(diAMsar)$^{3+}$ were employed as templates in both AlPO and GaPO synthesis gels spanning a wide range of M:P, temperature and pH. We have found that the diaminosarcophagines can be used at 190°C in AlPO preparations without significant decompositon. Our work has resulted in the synthesis of a number of new AlPO and GaPO materials which are presently being characterized. We have also started to use sarcophagine templates in aluminosilciate syntheses. The addition of cobalt(III) diaminosarcophagine salts to zeolite X synthesis gels results in samples that apparently have the intact complex trapped in their cavity systems. Under these high pH conditions the complex does not decompose significantly at temperatures less than $130^{\circ}C$ over a 48 hour period.

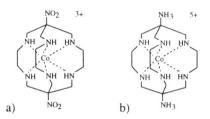

Figure 2. Cobalt (III) a) diNOsar and b) diAMsarH$_2$ complexes.

Synthesis of Aluminophosphate and Gallophosphate Materials. Typically, we prepared our aluminophosphate and gallophosphate materials by heating aqueous gels at 100 - 200°C (depending on the template and the need for large crystallites or high product purity) over periods of a few hours to a few days. The gels were prepared by the addition of aqueous phosphoric acid to aqueous suspensions of Vista Catapal B (pseudoboehmite ~80% Al$_2$O$_3$) in the case of AlPOs or "gallium oxyhydroxide" in the case of GaPOs. After aging, the gel pH was typically adjusted using aqueous TMAOH (tetramethylammonium hydroxide). The template species was then added to the gel as a halide salt, although in some cases we employed hydrogen phosphate salts. We have used a number of different "gallium oxyhydroxide" sources, including material prepared by the reaction of gallium with water at high temperatures, the products from the thermal decomposition of gallium nitrate, and the precipitate formed by the addition of ammonia to aqueous solutions of gallium chloride.

A few months after Bruce(43) reported our early work on compounds **1**, **6** and **7** in his Ph.D. thesis, Morgan and co-workers(32) also published a synthesis and structure for compound **1**. Our approach to the synthesis of this material, and the related compounds listed in Table I, is different from that of Morgan et al. We added the halide salt of the metal complex to an aluminophosphate gel along with TMAOH to adjust the pH. This avoids the preparation of Co(en)$_3$(OH)$_3$ from the halide by ion exchange chromatography or treatment with expensive silver compounds. However, there may be cases when the use of a halide salt along with TMAOH, rather than the hydroxide salt of the complex, will lead to different metal phosphate products.

Structures of the Aluminophosphate and Gallophosphate Materials.

The structural information that is necessary to evaluate both the chirality of a material and the nature of the template framework interactions has been obtained by diffraction techniques. In many cases it proved difficult to prepare crystals of sufficient size for characterization using in house X-ray equipment. However, it is possible using area detectors and high brightness synchrotron X-ray sources to structurally characterize extremely small crystals. Recent studies of ~30 x ~30 x 5 μm crystals of bacteriorhodopsin(57) and < 1μm³ crystals of kaolinite(58) illustrate the potential of this approach. We have employed synchrotron radiation to obtain structural data for several materials prepared using chelate complex templates (**3**, **8**, **10** and **11**).

Many of the materials that we have prepared are layered (see Table I) and can be though of as cationic templates sandwiched between aluminophosphate macroanionic sheets (see Figure 3). These materials contain water in the interlamellar space along with the chelate complexes.

Figure 3. $M(en)_3^{3+}$ in the interlayer space of d,l-$M(en)_3[Al_3P_4O_{16}]\cdot xH_2O$.

Although a number of different aluminophosphate layer compositions have been previously reported ($[Al_3P_4O_{16}]^{3-}$, $[Al_2P_3O_{12}]^{3-}$(59-61) and $[AlP_2O_8]^{3-}$(33)), so far, we have only prepared compounds containing $[Al_3P_4O_{16}]^{3-}$ macroanionic sheets. There are reports of ~15 different materials containing layers with this composition (see Table II), spanning five distinct types of layer structure. Materials **1-3**, **6-8** and **10** include representatives of three different types of $[Al_3P_4O_{16}]^{3-}$ structure (see Figure 4). Only one of the five known layer structures (**C**) has been prepared using both metal complexes (**6**) and organoammines as templates.(62-64)

Layer types **A** and **B** have only been prepared in the presence of metal complexes. Their connectivity is the same, but their structures are distinct due to the disposition of the terminal P=O groups relative to the plane of the layers. AlPO **1** contains achiral layers of type **A**, whereas materials **3**, **7**, and **8** all contain chiral sheet anions of type **B**. It should be noted that only one enantiomer of this chiral sheet anion is found in a given crystal of the materials (**3** and **7**) that contain resolved template ion, whereas both enantiomers of the sheet occur in every crystal when the achiral species s-cis-$Co(dien)_2^{3+}$ is used as a template. This indicates that chirality can be transferred from a chelate complex to a growing metal oxide framework. Additionally, all of the layered materials have only one enantiomer of the complex at any given position in the structure, indicating that the cation sites in a material are capable of distinguishing between enantiomers (there is chiral recognition).

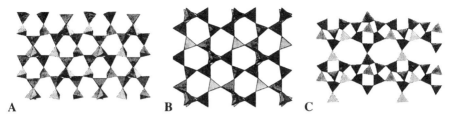

Figure 4. The three different $[Al_3P_4O_{16}]^{3-}$ layer types made using metal complexes.

Compound **4,** produced as a minority product during a failed attempt to synthesize **3**, contains discrete $[H_6Al_2P_6O_{24}]^{3-}$ clusters. Species of this type had never been prepared before, but can be considered to be a building block or fragment of the $[H_nAlP_2O_8]^{(3-n)-}$ chain anions reported by Jones,[65] Tieli *et al.*,[66] and recently prepared by us in the presence of $Co(en)_3^{3+}$[67] (see Figure 5). This cluster is also found as a fragment in the layered material $[NH_4]_3[Co(NH_3)_6]_3[Al_2(PO_4)_4]_2$.[33] While there has been very little work on aluminophosphate clusters, Riou and co-workers[68] have previously reported a material containing the species $[Al(PO_4)_4]^{9-}$ which can be regarded as a fragment of $[H_6Al_2P_6O_{24}]^{3-}$ and most other aluminophosphates.

Figure 5. The a) $[H_nAlP_2O_8]^{(3-n)-}$ chain, and b) $[H_6Al_2P_6O_{24}]^{3-}$ cluster structures

Our work with gallophosphates has been less extensive than that with aluminophosphates. However, we have prepared both a material (**2**) that is isostructural with layered aluminophosphate **1**, and a novel chiral 3D framework material **5**. The latter is the first structurally characterized 3D framework compound that we have prepared using chiral chelate complexes as templates. It crystallizes with a chiral interrupted framework and has a backbone connectivity closely related to that of diamond (see Figure 6). The framework is chiral because of the orientation of the terminal P=O and P-OH groups inside the material's pores. The pores are occupied by the chiral template cation which interacts with the framework via hydrogen bonds.

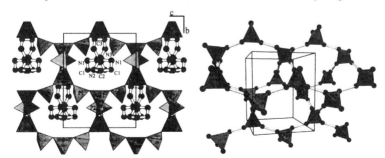

Figure 6. $[\Lambda-Co(en)_3][H_3Ga_2P_4O_{16}]$ the template location and the connectivity of the framework backbone (Reproduced from ref. *49*. Copyright 1997 ACS).

The charge densities of the known $Al_3P_4O_{16}^{3-}$ layer structures are given in Table II. Layer types **A**, **B**, **C**, and **D** have different charge densities, whereas layer type **E** has a charge density indistinguishable from that of type **D**. We believe that there are at least two important factors determining the type of layer structure that is formed in any particular synthesis; template charge to volume ratio and specific layer-template interactions. The layer is constrained to have a charge density within a range whose limits are related to the charge and size of the template species. The upper bound of this range is determined by the maximum number of templates that can be packed in a given area. The lower limit is harder to quantify as the packing density of the templates in the interlayer space can be lowered by changing the orientation of the template ions and by diluting them with uncharged species such as solvent molecules. In all of the $Al_3P_4O_{16}^{3-}$ structures that we are aware of there is extensive hydrogen bonding between the template species and the AlPO layer, this kind of specific interaction is presumably responsible for the ability of a given template to induce the growth of a particular layer type when two types with similar charge density exist.

We have begun to explore the proposed relationship between template charge to volume ratio and layer structure by performing syntheses with complexes that have both higher and lower charge to volume ratios than that of $Co(en)_3^{3+}$. Initial experiments have been performed using $Co(pn)_3^{3+}$, $Co(chxn)_3^{3+}$, and $Pt(en)_3^{4+}$ templates in AlPO syntheses. In each case crystalline AlPOs containing the metal complex were formed. Microcrystals of the products prepared using $Co(pn)_3^{3+}$ and $Pt(en)_3^{4+}$ have been examined using synchrotron radiation. Elemental analyses along with the unit cell parameters for these materials suggest that they are layered, but full structural characterization will require superior diffraction data.

Modification of Layered Aluminophosphates.

The layered AlPOs that we have prepared are structurally related to the clay minerals and members of the α-zirconium phosphate family. They have layer charge densities (4 -5 e/100Å^2) intermediate between those found for clay minerals*(69)* (1 -2.7 e/100Å^2 for smectites and ~4.4 for mica e/100Å^2) and α-zirconium phosphate*(70)* (~8.3 e/100Å^2). If the chelate complexes in the chiral layered AlPOs could be ion exchanged for other chiral ions, then it might be possible to use these materials for enantioselective ion exchange based separations. However, ion exchange of these materials is complicated by the AlPO layers susceptibility to hydrolysis. Our attempts at ion exchange using the achiral host d,l-$Co(en)_3[Al_3P_4O_{16}]$·xH_2O*(32,43)* in phosphate containing media (used to suppress hydrolysis) led to a completely new AlPO product d,l-$Co(en)_3[AlP_2O_8]$·xH_2O rather than an ion exchanged variant of the initial layer structure.*(67)*

Table II. Known Materials with $Al_3P_4O_{16}^{3-}$ Layers.

Template(s)	a/Å α/°	b/Å β/°	c/Å γ/°	SG	Stacking sequence	Layer spacing	Layer type	Charge e/100Å²	Ref
d,l-Co(en)$_3^{3+}$	8.561 90	21.323 90	13.813 90	Pnna	ABAB	10.66	A	5.07	(31,43)
d,l-Ir(en)$_3^{3+}$	8.550 90	21.931 90	13.936 90	Pnna	ABAB	10.97	A	5.03	(71)
s-cis-Co(dien)$_2^{3+}$	14.535 90	8.453 109.21	21.827 90	C2/c	ABAB	10.31	B	4.88	(50)
d-Co(en)$_3^{3+}$	8.503 90	14.619 90	20.88 90	C222$_1$	ABAB	10.44	B	4.85	(45)
d-$trans$-Co(dien)$_2^{3+}$	8.4575 90	8.4575 90	63.274 120	P6$_5$22	ABCDEF	10.55	B	4.84	(49)
d-Co(tn)$_3^{3+}$	8.862 90	14.706 108.87	11.402 90	P2$_1$	AAAA	11.402	C	4.60	(48)
(NH$_3$(CH$_2$)$_5$NH$_3$)$^{2+}$ C$_5$H$_{10}$NH^{2+}	9.801 90	14.837 105.65	17.815 90	P2$_{1/c}$	AAAA	9.44	C	4.54	(63,72)
EtNH$_3^+$	8.920 90	14.895 106.07	9.363 90	P2$_1$/n	AAAA	9.00	C	4.52	(62)
(NH$_3$(CH$_2$)$_2$NH$_3$)$^{2+}$ (OH(CH$_2$)OH$_2$)$^+$	9.014 90	14.771 90	17.704 90	P2$_1$2$_1$2$_1$	ABAB	8.85	C	4.51	(64)
PrNH$_3^+$	9.021 90	14.845 105.23	9.592 90	?	AAAA	9.26	C	4.48	(63)
[1,4-(NH$_3$)$_2$C$_6$H$_{10}$]$^{2+}$	12.937 90	12.937 90	18.246 120	P-3c1	ABAB	9.12	D	4.14	(73)
(NH$_3$(CH$_2$)$_4$NH$_3$)$^{2+}$	12.957 90	12.957 90	18.413 120	P-3c1	ABAB	9.21	D	4.13	(74)
Et$_3$NH$^+$	13.092 90	13.092 90	10.093 120	P3	AAAA	10.09	D	4.04	(75,76)
(NH$_3$CHMeCH$_2$NH$_3$)$^{2+}$	18.119 90	16.236 91.35	14.736 90	Ia		7.37	E	4.08	(77)

The Role of Hydrogen Bonding in Template Recognition and Dehydration.

In all of the AlPO materials that we have prepared there is extensive hydrogen bonding between the N-H groups on the complexes and both framework oxygens (primarily terminal P-O) and water. We believe that these interactions may be important in determining not only the dehydration behavior of the material after it is made but also the formation of a particular AlPO structure during the hydrothermal synthesis.

The formation of a given AlPO structure in the presence of a metal complex template is influenced by several factors. As mentioned earlier, charge balance must be achieved, and this will place constraints on the type of metal phosphate structure that can be formed. Additionally, hydrogen bonding between the template species and the framework as well as the space filling requirements of the template need to be considered. In examining the structures of the layered AlPOs that we have made, it becomes evident that for all of the materials prepared in the presence of a racemic mixture of the template species, there is only a single enantiomer of the template at a given cation site in the interlayer space. This implies that the growing structures are capable of distinguishing between the two enantiomeric forms of the template. This type of template chiral recognition may manifest itself as an ordered array of two enantiomers in the interlayer space or as the presence of only one enantiomer in a given AlPO crystal even though both enantiomers were present in the synthesis gel. The template recognition process may be a consequence of packing interactions between template species in the interlayer space, or template-AlPO interactions, or a combination of the two. It is notable that the cluster and chain compounds d,l-Co(en)$_3$[H$_6$Al$_2$P$_6$O$_{24}$] and d,l-Co(en)$_3$[AlP$_2$O$_8$]·xH$_2$O do not display good template recognition characteristics. In both materials the two enantiomers of the complex are disordered over the available cations sites. This disorder has been modeled for both materials. In the case of d,l-Co(en)$_3$[AlP$_2$O$_8$]·xH$_2$O the two enantiomers on a given site are oriented in such a way that they fill approximately the same 'pocket' in the crystal structure and the hydrogen bonding interactions (both in number and strength) between the AlPO chain and the N-H groups on the complex are essentially the same for both enantiomers. The primary hydrogen bonding interaction in this chain AlPO material is between a terminal P-O group located on the pseudo three fold axis of the template and three N-H groups on the triangular face of the complex that is perpendicular to this axis. This type of interaction is clearly not capable of distinguishing between enantiomers. The only other significant AlPO-template H-bonding interaction is also incapable of distinguishing between enantiomers. In the layered materials and the 3D framework GaPO that we have made, all of which have good template recognition characteristics, there are a multitude of H-bonding interactions between the framework and N-H groups on the complex. For example, in [Λ-Co(en)$_3$][H$_3$Ga$_2$P$_4$O$_{16}$] all 12 hydrogens on the N-H groups are within 2.45Å of a framework oxygen.

We have previously rationalized the observed weight loss versus temperature data for d-Co(en)$_3$[Al$_3$P$_4$O$_{16}$]·xH$_2$O(45) by considering the hydrogen bonding between interlayer water molecules and both the framework oxygens and the N-H groups on the template species. In the hydrated material there are two crystallographically distinct water sites with relative multiplicities 2:1. Water molecules on the higher multiplicty site are only hydrogen bonded to framework oxygens, whereas those on the lower multiplicity site interact with both N-H groups on the complex and framework oxygens. This difference in hydrogen bonding is consistent with the observed water loss which occurs in two stages; an initial 5.5% weight loss followed by a higher temperature 3% weight loss. Complex dehydration behavior has also been observed for d,l-Co(en)$_3$[Al$_3$P$_4$O$_{16}$]·xH$_2$O. In this case, *in-situ* powder X-ray diffraction was used to monitor the structural consequences of water loss.

An *In-Situ* Examination of Dehydration for d,l-Co(en)$_3$[Al$_3$P$_4$O$_{16}$]·xH$_2$O. The structural consequences of dehydrating d,l-Co(en)$_3$[Al$_3$P$_4$O$_{16}$]·xH$_2$O have been investigated by both ^{31}P NMR(43) and *in-situ* powder X-ray diffraction.(46) The X-ray experiments were performed using a scanning imaging plate system previously described by Norby.(78) Dry nitrogen was passed through capillary tubes containing the AlPO material and X-ray diffraction data were recorded as a function of time while the samples were heated (see Figure 7). The interpretation of the data is complicated by the difficulty of reproducing factors such as sample packing and nitrogen flow rate for every experiment. However, it appears that the dehydration of this AlPO occurs in at least two stages. On slow heating water loss initially leads to a decrease in the interlayer separation but no major structural changes. Under forcing conditions there is a major structural change that we believe probably involves a reorientation of the complexes in the interlayer space.

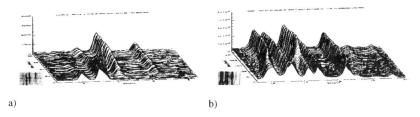

a)   b)

Figure 7. Sections of the powder XRD data for $[d,l\text{-Co(en)}_3][Al_3P_4O_{16}]\cdot xH_2O$, a) as it is isothermally dehydrated at $160°$ C, and b) as it is slowly heated to $250°$ C.

Evaluating Optical Purity.

Experimentally determining the optical purity of a bulk chiral solid sample is fraught with difficulties. Powder X-ray diffraction data is not capable of distinguishing between the enantiomers of a chiral solid. The use of optical measurements to quantitatively determine the optical purity of bulk samples is problematic because transparent disks containing the powdered material are needed and the interpretation of a measurement requires a knowledge of both the properties of an optically pure sample (including the anisotropy of these properties) and any texture present in the sample (powdered samples may display preferred orientation when pressed into a disk). In principle, the determination of the absolute configuration of many single crystals using diffraction techniques can provide an estimate of the optical purity of a sample. This method has the advantage that it can distinguish between samples that contain a mixture of crystals related to one another as enantiomers and crystals that are racemically twinned on some length scale. However, this approach requires large amounts of diffractometer time.

On two occasions we have tried to estimate the optical purity of chiral solid samples that we have prepared. In both cases we assumed that the optical purity of the template species was directly related to the purity of the chiral solid product. In the case of $[\Lambda\text{-Co(en)}_3][H_3Ga_2P_4O_{16}]$(47) we determined the optical purity of the residual template in the liquor from which the material was grown. Even though we do not usually have residual template in the mother liquor, this approach can be adapted to other systems by digesting the sample in acid prior to the optical measurements. However, the examination of template purity by optical methods is complicated by the dependence of the optical properties on the anions present in solution. For d-$Co(en)_3[Al_3P_4O_{16}]\cdot xH_2O$(45) we were able to estimate the optical purity of samples by powder diffraction. For this particular system, the presence of both enantiomer of the template in the synthesis gel leads to the formation of $d,l\text{-Co(en)}_3[Al_3P_4O_{16}]\cdot xH_2O$ rather than just a mixture of d-$Co(en)_3[Al_3P_4O_{16}]\cdot xH_2O$ and l-$Co(en)_3[Al_3P_4O_{16}]\cdot xH_2O$. $d,l\text{-Co(en)}_3[Al_3P_4O_{16}]\cdot xH_2O$ is distinguishable from $d(\,or\,l)$-$Co(en)_3[Al_3P_4O_{16}]\cdot xH_2O$ by powder diffraction as they are not related as enantiomers.

50

Conclusions

Chelate complexes constitute a versatile family of template species. Chirality can be transferred from an optically pure complex to a growing inorganic framework under appropriate conditions. Template species that are stable under a wide variety of framework synthesis conditions can be prepared if the ligand system and metal ion are chosen carefully. However, the preparation of optically pure chiral frameworks with microporosity that is accessible to small organic molecules is still a significant challenge. We are presently developing strategies for the synthesis of such materials.

Acknowledgments

Contributions to the above work were made by D. A. Bruce, M. J. Gray, S. M. Stalder, J. C. Jasper, J. Kruger, D. J. Williams, M. G. White and J. A. Bertrand. I am grateful to the donors of the Petroleum Research Fund, administered by the American Chemical Society, to the Hoechst-Celanese Corporation, and to the Molecular Design Institute at the Georgia Institute of Technology, supported under ONR Contract N00014-95-1-1116, for funding this work. The synchrotron experiments were performed at beam line X7B, NSLS, Brookhaven. This facility was supported under contract DE-AC02-76CH00016 with the U.S. Department of Energy by its division of Chemical Sciences, Office of Basic Research. I am grateful to J. C. Hanson and P. Norby for assistance while performing experiments at the NSLS.

Literature Cited

1)Stinson, S. C. *Chem. Eng. News* **1997**, *October 20*, 38-70.
2)Davis, M. E.; Lobo, R. F. *Chem. Mater.* **1992**, *4*, 756-768.
3)Treacy, M. M. J.; Newsam, J. M. *Nature* **1988**, *332*, 249-251.
4)Newsam, J. M.; Treacy, M. M. J.; Koetsier, W. T.; de Gruyter, C. B. *Proc. Roy. Soc. (London) A* **1988**, *420*, 375-405.
5)Higgins, J. B.; LaPierre, R. B.; Schlenker, J. L.; Rohrman, A. C.; Wood, J. D.; Kerr, G. T.; Rohrbaugh, W. J. *Zeolites* **1988**, *8*, 446-452.
6)Davis, M. E. *Acc. Chem. Res.* **1993**, *26*, 111-115.
7)Anderson, M. W.; Terasaki, O.; Ohsuna, T.; Philippou, A.; MacKay, S. P.; Ferreira, A.; Rocha, J.; Lidin, S. *Nature* **1994**, *367*, 347-351.
8)Rouse, R. C.; Peacor, D. R. *Am. Mineral.* **1986**, *71*, 1494-1501.
9)Harrison, W. T. A.; Gier, T. E.; Stucky, G. D.; Broach, R. W.; Bedard, R. A. *Chem. Mater.* **1996**, *8*, 145-151.
10)Feng, P.; Bu, X.; Tolbert, S. H.; Stucky, G. D. *J. Am. Chem. Soc.* **1997**, *119*, 2497-2504.
11)Hutchings, G. J.; Willock, D. J. *212th ACS meeting, Orlando, Florida, August 1996* **1996**, *Abstract CATL 008*.
12)Feast, S.; Bethell, D.; Bulman Page, P. C.; King, F.; Rochester, C. H.; Siddiqui, M. R. H.; Willock, D. J.; Hutchings, G. J. *J. Chem. Soc. Chem. Commun.* **1995**, 2409-2411.
13)Willock, D. J.; Bethell, D.; feast, S.; Hutchings, G. J.; King, F.; Page, P. C. *Top. Catal.* **1996**, *3*, 77-89.
14)Feast, S.; Bethell, D.; Page, P. C.; Siddiqui, M. R.; Willock, D. J.; Hutchings, G. J.; King, F.; Rochester, C. H. *Stud. Surf. Sci. Catal.* **1996**, *101*, 211-219.
15)Feast, S.; Siddiqui, M. R. H.; Wells, R. P. K.; Willock, D. J.; King, F.; Rochester, C. H.; Bethell, D.; Page, P. C. B.; Hutchings, G. J. *J. Catal.* **1997**, *167*, 533-542.

16)Feast, S.; Bethell, D.; Page, P. C. B.; King, F.; Rochester, C. H.; Siddiqui, M. R. H.; Willock, D. J.; Hutchings, G. J. *J. Mol. Catal. A:Chem.* **1996**, *107*, 291-295.

17)Willock, D. J.; Bethell, D.; Feast, S.; Hutchings, G. J.; King, F.; Page, P. C. B. *Top. Catal.* **1996**, *3*, 77-89.

18)Hutchings, G. J.; Wells, R.; Feast, S.; Siddiqui, M. R. H.; Willock, D. J.; King, F.; Rochester, C. H.; Bethhell, D.; Page, P. C. B. *Cat. Lett.* **1997**, *46*, 249-254.

19)Leibovitch, M.; Olovsson, G.; Sundarababu, G.; Ramamurthy, V.; Scheffer, J. R.; Trotter, J. *J. Am. Chem. Soc.* **1996**, *118*, 1219-1220.

20)Sundarababu, G.; Leibovitch, M.; Corbin, D. R.; Scheffer, J. R.; Ramamurthy, V. *Chem. Commun.* **1996**, 2159-2160.

21)Ogunwumi, S. B.; Bein, T. *Chem. Commun.* **1997**, 901-902.

22)Sabater, M. J.; Corma, A.; Domenach, A.; Fornes, V.; Garcia, H. *Chem Commun.* **1997**, 1285-1286.

23)Lok, B. M.; Cannan, T. R.; Messina, C. A. *Zeolites* **1983**, *3*, 282-291.

24)Gilson, J.-P. In *Zeolite Microporous Solids: Synthesis, Structure, and Reactivity*; Derouane, E. G., Lemos, F., Naccache, C. and Ribeiro, F. R., Eds.; Kluwer: Dordrecht, 1992, pp 19-48.

25)Guth, J. L.; Caullet, P.; Seive, A.; Patarin, J.; Delprato, F. In *Guidelines for Mastering the Properties of Molecular Sieves*; Barthomeuf, D., Ed.; Plenum Press: New York, 1990, pp 69-85.

26)Balkus, K. J.; Hargis, C. D.; Kowalak, S. *ACS Symp. Ser.* **1992**, *499*, 347-354.

27)Balkus, K. J.; Kowalak, S.; Ly, K. T.; Hargis, D. C. *Stud. Surf. Sci. Catal.* **1991**, *69*, 93-99.

28)Balkus, K. J.; Gabrielov, A. G.; Bell, S. L.; Bedioui, F.; Roue, L.; Devynck, J. *Inorg. Chem.* **1994**, *33*, 67-72.

29)Kowalak, S.; Balkus, K. J. *Collect. Czech. Chem. Commun.* **1992**, *57*, 774-780.

30)Rankel, L. A.; Valyocsik, E. W. *U.S. Patent* **1985**, *4,500,503*.

31)Bruce, D. A.; Bertrand, J. A.; Occelli, M. L.; White, M. G.; Mertens, F. In *Catalysis of Organic Reactions*; Scaros, M. G. and Prunier, M. L., Ed.; Marcel Dekker, Inc., 1995, pp 545-551.

32)Morgan, K.; Gainsford, G.; Milestone, N. *J. Chem. Soc. Chem. Commun.* **1995**, 425-426.

33)Morgan, K. R.; Gainsford, G. J.; Milestone, N. B. *Chem Commun.* **1997**, 61-62.

34)Valyocsik, E. W. *U.S. Patent* **1986**, *4,568,654*.

35)Balkus, K. J.; Shepelev, S. *Prepr. Pap. Am. Chem. Soc., Div. Petr.* **1993**, *38*, 512-515.

36)Balkus, K. J.; Gavrielov, A. G.; Sandler, N.; Jacob, T. *North American Catalysis Society, Snowbird, Utah* **1995**, *Meeting Abstracts*, T115-T116.

37)Balkus, K. J.; Gabrielov, A. G.; Zones, S. I. *Stud. Surf. Sci. Catal.* **1995**, *97*, 519-525.

38)Freyhardt, C. C.; Tsapatsis, M.; Lobo, R. F.; Balkus, K. J.; Davis, M. F. *Nature* **1996**, *381*, 295-298.

39)Balkus, K. J.; Kowalak, S. *U.S. Patent* **1992**, *5,167,942*.

40)Keene, F. R.; Searle, G. H.; Yoshikawa, Y.; Imai, A.; Yamasaki, K. *J. Chem. Soc. Chem. Commun.* **1970**, 784-786.

41)Keene, F. R.; Searle, G. H. *Inorg. Chem.* **1972**, *11*, 148-156.

42)Searle, G. H.; Keene, F. R. *Inorg. Chem.* **1972**, *11*, 1006-1011.

43)Bruce, D. A. *PhD. dissertation*, Georgia Institute of Technology: Atlanta, 1994.

44)Wilkinson, A. P.; Gray, M. J.; Stalder, S. M. *Mat. Res. Soc. Symp. Proc.* **1996**, *431*, 21-26.

45)Gray, M. J.; Jasper, J.; Wilkinson, A. P.; Hanson, J. C. *Chem. Mater.* **1997**, *9*, 976-980.

46)Gray, M. J. *MS dissertation*; Georgia Institute of Technology: Atlanta, 1997.

47)Stalder, S. M.; Wilkinson, A. P. *Chem. Mater.* **1997**, *9*, 2168-2173.

48)Bruce, D. A.; Wilkinson, A. P.; White, M. G.; Bertrand, J. A. *J. Chem. Soc. Chem. Commun.* **1995**, 2059-2060.
49)Bruce, D. A.; Wilkinson, A. P.; White, M. G.; Bertrand, J. A. *J. Solid State Chem.* **1996**, *125*, 228-233.
50)Kruger, J.; Wilkinson, A. P., unpublished results.
51)Galsbol, F.; Rasmussen, B. S. *Acta Chem. Scand.* **1982**, *A 36*, 83-87.
52)Galsbol, F. *Proc. 16th Int. Conf. Coord. Chem.* Univeristy College, Dublin: Dublin, 1974, pp R28.
53)Galsbol, F. *Acta Chem. Scand.* **1978**, *A 32*, 757-761.
54)Creaser, I. I.; Geue, R. J.; Harrowfield, J. M. B.; Herlt, A. J.; Sargeson, A. M.; Snow, M. R.; Springborg, J. *J. Am. Chem. Soc.* **1982**, *104*, 6016-6025.
55)Geue, R. J.; McCarthy, M. G.; Sargeson, A. M. *J. Am. Chem. Soc.* **1984**, *106*, 8282-8291.
56)Harrowfield, J. M. B.; Lawrance, G. A.; Sargeson, A. M. *J. Chem. Ed.* **1985**, *62*, 804-806.
57)Pebay-Peyroula, E.; Rummel, G.; Rosenbusch, J. P.; Landau, E. M. *Science* **1997**, *277*, 1676-1681.
58)Neder, R. B.; Burghammer, M.; grasl, T.; Schulz, H.; Bram, A.; Fiedler, S.; Riekel, C. Z. *Kryst.* **1996**, *211*, 763-765.
59)Chippindale, A. M.; Powell, A. V.; Bull, L. M.; Jones, R. H.; Cheetham, A. K.; Thomas, J. M.; Xu, R. *J. Solid State Chem.* **1992**, *96*, 199-210.
60)Oliver, S.; Kuperman, A.; Lough, A.; Ozin, G. A. *Chem. Mater.* **1996**, *8*, 2391-2398.
61)Oliver, S.; Kuperman, A.; Lough, A.; Ozin, G. A. *J. Chem. Soc. Chem. Commun.* **1996**, 1761-1762.
62)Gao, Q.; Li, B.; Chen, J.; Li, S.; Xu, R.; Williams, I.; Zheng, J.; Barber, D. *J. Solid State Chem.* **1997**, *129*, 37-44.
63)Chippindale, A. M.; Natarajan, S.; Thomas, J. M.; Jones, R. H. *J. Solid State Chem.* **1994**, *111*, 18-25.
64)Jones, R. H.; Thomas, J. M.; Xu, R.; Huo, Q.; Cheetham, A. K.; Powell, A. V. *J. Chem. Soc. Chem. Commun.* **1991**, 1266-1268.
65)Jones, R. H.; Thomas, J. M.; Xu, R.; Huo, Q.; Xu, Y.; Cheetham, A. K.; Bieber, D. *J. Chem. Soc. Chem. Commun.* **1990**, 1170-1172.
66)Tieli, W.; Long, Y.; Wenqin, P. *J. Solid State Chem.* **1990**, *89*, 392-395.
67)Jasper, J. D.; Wilkinson, A. P. *Chem. Mater.* **1997**, *10*, 1664-1667.
68)Riou, D.; Loiseau, T.; Ferey, G. *J. Solid State Chem.* **1992**, *99*, 414-418.
69)Pinnavaia, T. J.; Hyungrok, K. In *Zeolite Microporous Solids: Synthesis, Structure, and Reactivity*; Derouane, E. G., Lemos, F., Naccache, C. and Ribeiro, F. R., Eds.; Kluwer, 1992, pp 79-90.
70)Alberti, G.; Costantino, U. *J. Mol. Catal.* **1984**, *27*, 235-250.
71)Williams, D.; Wilkinson, A. P. unpublished results.
72)Jones, R. H.; Chippindale, A. M.; Natarajan, S.; Thomas, J. M. *J. Chem. Soc. Chem. Commun.* **1994**, 565-566.
73)Barrett, P. A.; Jones, R. H. *J. Chem. Soc. Chem. Commun.* **1995**, 1979-1981.
74)Thomas, J. M.; Jones, R. H.; Xu, R.; Chen, J.; Chippindale, A. M.; Natarajan, S.; Cheetham, A. K. *J. Chem. Soc. Chem. Commun.* **1992**, 929-931.
75)Oliver, S.; Kuperman, A.; Lough, A.; Ozin, G. A.; Garces, J. M.; Olken, M. M.; Rudolf, P. *Stud. Surf. Sci. Catal.* **1994**, *84*, 219-225.
76)Kuperman, A.; Nadimi, S.; Oliver, S.; Ozin, G. A.; Garces, J. M.; Olken, M. M. *Nature* **1993**, *365*, 239-242.
77)Williams, I. D.; Gao, Q.; Chen, J.; Ngai, L.-Y.; Lin, Z.; Xu, R. *J. Chem. Soc. Chem. Commun.* **1996**, 1781-1782.
78)Norby, P. *Mater. Sci. For.* **1996**, *228-231*, 147-152.

Chapter 5

The Design, Synthesis, and Characterization of Redox-Recyclable Materials for Efficient Extraction of Heavy Element Ions from Aqueous Waste Streams

Peter K. Dorhout and Steven H. Strauss

Department of Chemistry, Colorado State University, Fort Collins, CO 80523

The design and synthesis of redox-recyclable materials for the extraction of heavy element ions from aqueous waste streams has involved the synthesis of activated extractants, the extraction of target ions and the recovery of those waste contaminants by an efficient cycle which minimizes secondary waste volume and provides efficient decontamination of the primary waste stream. The feasibility of using lithium-intercalated transition metal disulfides, Li_xMS_2 as redox-recyclable ion-exchange materials for the extraction of the heavy metal ions Hg(II), Pb(II), Cd(II), Zn(II), Au(III), Ag(I), and Cu(I) was investigated ($0.25 \leq x \leq 1.9$, M = Mo, W, Ti, Ta). The materials Li_xMoS_2 and Li_xWS_2 removed significant amounts of the heavy metal ions in question, in one case from 200 ppm aqueous Hg down to ~5 ppb Hg, yielding ion-exchanged materials such as $Hg_{0.5}MoS_2$ and $Pb_{0.15}MoS_2$. The selectivity of Li_xMoS_2 for selected heavy-metal ions was Hg(II) > Pb(II) > Cd(II) > Zn(II). Heating solid Hg_yMoS_2 under vacuum at 425 °C initiated an entropy-driven internal redox reaction which resulted in the deactivation of the extractant, producing essentially mercury-free MoS_2 and mercury vapor. Samples of Li_xMoS_2 were used for three complete activation-extraction-deactivation(recovery) cycles with no loss in ion-exchange capacity. The maximum capacity for Hg was 580 mg Hg/g extractant when $Li_{1.9}MoS_2$ was used. For Li_xTiS_2 and Li_xTaS_2, hydrolysis produced $S^{2-}_{(aq)}$ ions which precipitated Hg(II) as $HgS_{(s)}$. Although these materials removed significant amounts of Hg(II), their hydrolysis precluded their use as redox-recyclable extractants.

There are numerous practical reasons for selectively separating *heavy metal ions* of all types from aqueous media. A few obvious examples are the remediation of hazardous or radioactive waste,[1] the remediation of contaminated groundwater,[2] and the recovery of precious and/or toxic metals from industrial processing solutions.[3] A variety of well-known techniques are available to the chemist or engineer for these tasks, including solvent extraction,[4] ion-exchange chromatography,[5] and precipitation.[6] In modern applications of these techniques, *the recovery and reuse* of the extractant materials are becoming more and more important. This is being driven by tougher environmental regulations,[7] the high initial costs of new, more effective, and more selective extractants, and the need to minimize the volume of secondary waste destined for permanent disposal.

Many heavy metal ions are "soft" Lewis acids, which means that their affinities for soft donor atoms such as phosphorus and sulfur are considerably higher than for hard donor atoms such as nitrogen and oxygen.[8] Examples of soft metal ions are Hg^{2+}, Pb^{2+}, Cd^{2+}, Cu^{2+}, Ag^+, Au^{3+}, Pt^{4+}, and Tl^+. (Examples of hard metal ions include Li^+ and the other alkali metal ions, Mg^{2+} and the other alkaline earth metal ions, Ti^{4+}, and Cr^{3+}.) Many of the soft metal ions are of concern because they are highly toxic and are present in a variety of waste streams that can potentially contaminate the environment if released. For example, the following soft metal ions are currently identified for regulation under RCRA/SDWA:[9] Cu^{2+}, Ag^+, Cd^{2+}, Hg^{2+}, Pb^{2+}, and Tl^+.[10] Safe, efficient, and cost effective separation *and recovery* of these metal ions from waste streams is therefore an important scientific and technological goal. The design and synthesis of materials that can meet these goals is an important focus for new materials synthesis.

Contamination or potential contamination of surface waters and groundwater has already occurred in many locations, necessitating, as above, safe, efficient, and cost effective remediation in the very near future. For example, the Department of Energy must pre-treat, treat, and dispose of a number of aqueous high-level and mixed waste streams that contain mercury, lead, and/or other soft heavy metals.[11,12] To satisfy *environmental concerns*, these soft, heavy elements must be removed from the waste (along with certain other heavy metals and radionuclides) before the bulk of the remaining waste can be disposed of in sub-surface vaults. As a particular case in point, there are currently 1.5×10^6 gallons of ICPP sodium-bearing waste in which $[H_3O^+]$ ~1.5 M, $[Hg^{2+}]$ ~0.002 M, and $[Pb^{2+}]$ ~0.001 M.[12] Mercury is present in these streams because it was added to the acidic processing solutions to catalyze the dissolution of aluminum metal. Current technologies for the selective extraction of mercury and lead need to be improved or replaced for the following reasons: (i) *selectivity* for mercury and lead should be improved; (ii) *decontamination factors* should be increased; (iii) *recovery* of mercury and lead from the extractants should be optimized to allow for recycling and reuse of the extractants, lower costs, and minimization of the volume and mass of the final form of mercury and lead. New materials must be created and existing materials must be modified in an effort to provide new technologies for heavy-element ion waste remediation.

Besides mercury- and lead-contaminated waste in the DOE complex, there are many other examples of anthropogenic or naturally-occuring soft-heavy-metal-ion pollution that are of significant environmental concern. For example, high concentrations of lead are found in soils in many U.S. urban areas.[13] Lead must also be removed from contaminated wastes and soil at battery recycling Superfund sites.[14] Both lead and copper contaminate soil at the Aberdeen Proving Ground in Maryland,[15] and lead, copper, and cadmium are present in storm water runoffs entering Lakes Bay, New Jersey.[16] Highly toxic thallium(I) is present in natural waters and waste waters.[17] Mercury, of course, is present in many aquatic environments, and the remediation of such environments, or the avoidance of contamination, is an area of active interest.[18]

Although our work is concerned with removing soft heavy metal ions from *aqueous* solutions, the work is relevant to the extraction of these metal ions from non-aqueous waste streams, contaminated slurries, and contaminated soils. Soil remediation, for example, involves leaching the metal ions out of the soil and into aqueous solutions (e.g. acetic acid/sodium acetate[19] or acidic potassium iodide[20]). Clearly, the development of methodologies for recovering heavy metal ions from aqueous solution is a desirable goal.

The volume of secondary waste is a major consideration for a remediation problem such as the decontamination of a waste stream containing, for example, $<10^{-3}$ M Hg(II). In general, waste management using separation technologies becomes more difficult and expensive as feed concentrations in the primary waste stream decrease. It has been shown that the volume of secondary waste produced per mole of pollutant recovered is inversely related to the feed concentration.[21]

The State-of-the-Art. Several recent, advanced methodologies for removing soft metal ions such as Hg^{2+} and Pb^{2+} from aqueous solutions have been reported. For example, one strategy has been to design chelating ligands that are specific for soft metal ions (these usually contain sulfur atoms). An example of one such study is the synthesis of Pb^{2+} complexes of thiohydroxamic acids.[22] In another study, sulfur-bearing calixarenes complexed Hg^{2+} and CH_3Hg^+ and allowed for the extraction of these ions from an aqueous phase into an organic phase.[23] Sulfur ligands are not an absolute requirement: ionizable crown ethers have been used to extract Hg^{2+} into supercritical carbon dioxide.[24] Cryptands and other polydentate ligands have been used to extract Tl^+ from aqueous phases into organic solvents.[25] Another strategy has been to use water soluble polymers, including polyethyleneimine, to complex metal ions such as Hg^{2+} and Pb^{2+} and to remove them from the aqueous phase by ultrafiltration.[26]

The strategy that is most relevant to our work is the use of solid-state materials as sorbents for soft heavy metal ions. Natural zeolites and zeoliferous rocks have been used to remove Hg^{2+}, Cd^{2+}, and Ag^+ from aqueous phases.[27] In addition, some silicates intercalated with various organic compounds have been shown to sorb metal ions such as Hg^{2+}, Pb^{2+}, Cd^{2+}, Cu^{2+}, and Ag^+.[28] Clays and related minerals have been extensively explored as heavy-metal-ion sorbents.[29] The interlamellar surfaces of Montmorillonite have been covalently functionalized with sulfur-containing moieties such as thiols, as shown in Figure 1. One of these composite materials, named thiomont, was shown to effectively remove Hg(II) and Pb(II) from aqueous media (capacities were 70 mg Pb(II)/g thiomont and 65 mg Hg(II)/g thiomont).[29] A significant finding of this study was that acid leaching could be used to regenerate the thiomont. The authors did not state whether the capacity of the regenerated material for mercury was the same or was lower than the original sample of thiomont, although they did report that the capacity for lead did not diminish after regeneration.

In another relevant study, the capacity for Hg(II) of mesoporous silica with thioalkyl groups grafted to the surface, as shown in Figure 2, was measured.[30] This solid had a capacity of 505 mg Hg/g of extractant. Recovery of mercury and partial regeneration of the extractant was accomplished by washing the mercury-loaded material with concentrated aqueous HCl. The capacity of the regenerated extractant, however, was only 40% of the original capacity. Although this material and thiomont have a high capacity for aqueous Hg(II), their regeneration results in the recovery of mercury in a highly acidic aqueous secondary waste stream. This may preclude their use for waste streams where a reduction of waste volume is essential.

The New Approach. Our approach uses an unexplored class of potential solid-state extractants, *redox-active layered (or channeled) metal chalcogenides*, some of which have been synthesized for the first time in the course of this work. This idea is based on a solvent extraction process that was initially used for the removal of $^{99}TcO_4^-$ from water.[31] The essential features of our idea, applied to ion exchange, are:

- some layered solids (hosts) *reversibly* intercalate metal ions (guests);
- some layered metal chalcogenides can be partially *reduced*, providing the driving force for the intercalation by charge-compensating metal ions;

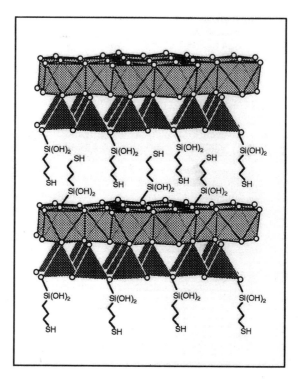

Figure 1   The inner surfaces of Montmorillonite functionalized with thiols
(Adapted from reference 29. Copyright 1995.).

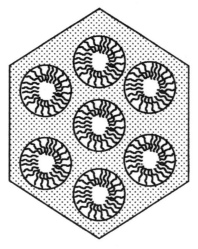

Figure 2   The inner channels of a mesoporous silica functionalized with thiols
(Adapted from reference 30. Copyright 1997.).

- the chalcogenides sulfur, selenium, and tellurium are soft donors, so the selectivity of reduced chalcogenide materials for *soft* charge-compensating metal ions (i.e., pollutants such as Hg^{2+}, Pb^{2+}, etc.) will be very high;
- *deactivation* of the host-guest material back to the original layered metal chalcogenide would provide for the *recovery* of pollutant metal ions and for the *reuse* of the solid-state extractant.

A potential activation (**1**), extraction (**2**), and deactivation/recovery (**3**) cycle is illustrated in Figure 3, which uses Li^+ as the original charge-compensating metal ion, Hg^{2+} as the pollutant, and chemical oxidation in the deactivation/recovery step for a layered solid example. The driving force for the extraction step in the cycle is provided by the interchange of hard Li^+ from the soft, chalcogen-rich environment to the hard, aqueous environment and soft Hg^{2+} from the hard aqueous environment to the soft, chalcogen-rich environment. This follows the well-documented principles of hard/soft Lewis acid/base theory.[8]

There are numerous scientific questions and technical problems that we address in the discussion below that explain our choice of extractants. First, the solid-state, layered or channeled, metal-chalcogenide extractants, in both active (reduced) and inactive (oxidized) forms, must be stable in contact with a wide variety of aqueous phases at a variety of pH values. Second, the solid-state extractants must have suitable electrochemical potentials, so that they can be shuttled between their active and inactive oxidation states using relatively mild and inexpensive reductants and oxidants. Third, solid-state extractant activation and deactivation (redox) reactions must be sufficiently rapid to allow for reasonably short time intervals for complete activation/extraction/recovery cycles. Fourth, the solid-state extractants must not dissolve during prolonged contact with whatever aqueous phase is to be treated. They must not undergo any other form of decomposition (including irreversible over-reduction or over-oxidation) during many cycles of activation, extraction, and recovery. Finally, practical solid-state extractants should have low toxicity, low cost, and should be relatively easy to prepare and handle.

The solids of choice are the layered binary metal dichalcogenides of the 2H-MoS_2 and 1T-TiS_2 structure types. Of the 13 known layered binary metal sulfides, two main structure types prevail.[32] These two main structure types are the CdI_2 and MoS_2 polytypes, shown below in Figure 4. In the CdI_2 polytype, the chalcogenides are arranged in an hexagonal close-packed layered structure and the metal atoms are found in octahedral coordination sites. The metal sites are arranged so that one-half of the available octahedral holes in the close-packed structure are filled in alternating layers that are then held together by relatively weak van der Waals forces. Metal sulfides that form in this structure type include SnS_2, TaS_2, TiS_2, and ZrS_2. In the MoS_2 polytype, the chalcogenides are arranged in a modified hexagonal close-packed layered array and the metal atoms are found in trigonal prismatic holes. This is accomplished by a chalcogenide-layer stacking pattern AABBAABB... In this polytype, the metal atoms occupy every second trigonal prismatic hole in the AA or BB layers, leaving the octahedral holes between A and B layers empty. As in the CdI_2 or 1T-polytype, the layers are held together by van der Waals forces. Some examples of compounds that crystallize in this type are MoS_2 and WS_2.

Intercalation of MQ_2 (Q = S, Se, Te) host solids like MoS_2 is not unique to our studies, although using them as selective ion-exchange resins is unique.[33] Morrison has used Li_xMS_2 as a source of exfoliated MS_2 charged layers for restacking of new homometallic and hetermetallic layered solids.[34] The lithium intercalated MQ_2 solids have had an interesting past as possible solid electrolytes in solid-state batteries.[35] Hibma provided an extensive review of monovalent cation intercalation into layered metal dichalcogenides.[36] In this review, Hibma details structural descriptions of a series of MQ_2 solids and their monovalent cation guest-host

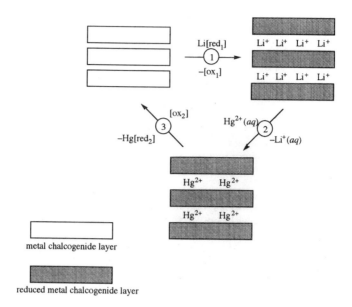

metal chalcogenide layer

reduced metal chalcogenide layer

Figure 3   A potential redox-recyclable scheme for the activation (1), extraction (2), and recovery/deactivation (3) of mercury using a layered metal chalcogenide extraction material.

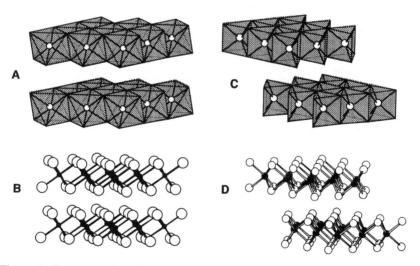

Figure 4   Two examples of metal chalcogenide redox-recyclable extractants used in our studies. (A) and (B) are two views of the CdI_2 structure where the metal atom is in an octahedral site. (C) and (D) are two views of the $2H\text{-}MoS_2$ structure where the metal atom is in a trigonal prismatic site.

counterparts. The phase relationships between the 2H-type and the smaller-cell 1T-type are described, and the effects of intercalative staging are discussed in detail. A more recent review from the *NATO ASI Series on the Chemical Physics of Intercalation* provides insight from Rouxel, Schöllhorn, Murphy, McKinnon, and Frindt on the chemistry and physics of intercalated metal chalcogenide host materials.[37] The known intercalation chemistry of MQ_2 compounds inspired us to consider these materials as *selective* ion-exchange host materials for the remediation of heavy element ions from aqueous waste.

Detailing New Directions.

We have initiated a study of the extraction of Hg^{2+}, Pb^{2+} and other metal ions from a variety of aqueous solutions using redox-active layered metal chalcogenides.[38] We began our investigation with MS_2 solids because they appeared to meet many of the design criteria highlighted earlier. Most are relatively inexpensive, nontoxic, very stable in both the deactivated, or oxidized form, and the activated, or reduced form and can be readily reduced with the concomitant intercalation of Li^+. Several systems are, however, subject to irreversible hydrolysis, and their decomposition will also be discussed. Finally, we have attempted to expand this project into the realm of shape-controlled synthesis of new solid extractants which generate very large surface area MS_2 solids that may be capable of being membrane-bound materials that may have extraordinary capacities for heavy metal ions.

We have focused our studies on very acidic waste streams and considered only those systems that have been notoriously difficult and very hostile towards other extraction methods in the past. Results of our studies demonstrate the viability of using lithium-intercalated molybdenum disulfide, Li_xMoS_2, as a solid-state extractant for soft metal ions. The three steps illustrated in Figure 3 can be summarized here specifically for Li_xMoS_2 but may be generalized for all MS_2 activated solids:

Activation $\quad MoS_{2(s)} + x\,[Li][red_1]_{(org)} \rightarrow Li_xMoS_{2(s)} + x\,[ox_1]_{(org)}$ (1)

Extraction: $\quad Li_xMoS_{2(s)} + x/2\,Hg^{2+}_{(aq\ waste)} \rightarrow Hg_{x/2}MoS_{2(s)} + x\,Li^+_{(aq\ waste)}$ (2)

Recovery/deactivation: $\quad Hg_{x/2}MoS_{2(s)} + x/2\,[ox_2] \rightarrow MoS_{2(s)} + x/2\,Hg[ox_2]$ (3)

Samples of Li_xMoS_2 prepared in our laboratory had the reproducible stoichiometry $x = 1.3$ (other stoichiometries are possible if the conditions in (1) are varied). Samples of this material were introduced into acidic aqueous solutions (0.1 M HNO_3 as a waste stream simulant) of different divalent cations, including Hg^{2+}, Pb^{2+}, Cd^{2+}, Zn^{2+}, Ba^{2+}, and Mg^{2+}. Some of the ion-exchange capacity of $Li_{1.3}MoS_2$ was lost due to hydrogen evolution when this material was treated with water, the amount quantified by Toeppler pump experiments on a calibrated vacuum line, equation (4):

$$Li_{1.3}MoS_{2(s)} + 1.3H_3O^+_{(aq)} \rightarrow$$
$$(H_3O)_{0.78}MoS_{2(s)} + 0.26H_{2(g)} + 1.3Li^+_{(aq)} + 0.52H_2O_{(l)}$$ (4)

There may also be some loss of capacity due to oxidation of $Li_{1.3}MoS_2$ by oxygen (the experiments were not carried out under anaerobic conditions). The exchange of intercalated Li^+ (and, consequently, H_3O^+) for the divalent metal ions (M^{2+}) was determined by measuring the aqueous concentrations of the divalent metal ions before and after treatment with solid $Li_{1.3}MoS_2$ and considering (4) above:

$$(H_3O)_{0.78}MoS_{2(s)} + z\,M^{2+}_{(aq)} \rightarrow$$
$$(H_3O)_{0.78-2z}M_zMoS_{2(s)} + 1.3\,Li^+_{(aq)} + 2z\,H_3O^+_{(aq)}$$ (5)

Our data are listed in Table I. Note that the theoretical maximum value of z is 0.65 (one-half of 1.3), a value that could only be reached if there was no hydrogen evolution, equation (4). Therefore, the minimum value of Mo/M^{2+}, or $1/z$, is 1.8.

Table I. Mo/M^{2+} Mole Ratios in $(H_3O)_{0.78-2z}M_zMoS_2$ Formed by Treatment of $Li_{1.3}MoS_2$ with Aqueous Solutions of M^{2+} Metal Ions.

M^{2+} Metal Ion	Mo/M^{2+} Mole Ratio
	1.8[a]
Hg^{2+}	2.0(2)
Pb^{2+}	11(1)
Cd^{2+}	22(1)
Zn^{2+}	40(2)
Mg^{2+}	63(1)

[a] The hypothetical minimum value (see text)

As discussed in the introduction, extractant capacity and selectivity are only two of the four important design criteria for modern extractants. The other two criteria are the feasibility of recovering the pollutant in a minimal volume of secondary waste and the feasibility of reusing the extractant once the pollutant has been recovered. Mercury could not be recovered from $Hg_{0.32}MoS_2$ by oxidation with O_2 but could be recovered by heating the host-guest solid under vacuum. When $Hg_{0.32}HgS_2$ was heated under vacuum to 425°C, the entropy-driven internal redox reaction depicted in equation (6) resulted in the formation of mercury vapor and polycrystalline MoS_2

$$Hg_{0.32}MoS_{2(s)} \xrightarrow{\quad 425°C \quad} 0.32\, Hg_{(g)} + MoS_{2(s)} \qquad (6)$$

(confirmed by powder X-ray diffraction, XRD). The Hg vapor was collected in a cold trap at –196°C. This procedure resulted in the recovery of >95% of the mercury originally present in $Hg_{0.32}MoS_2$ as elemental mercury. Only a trace amount of mercury was detected by ICP-AES when the recovered MoS_2 was digested in *aqua regia* (specifically, the stoichiometry of the recovered compound was $Hg_{0.02}MoS_2$). Elemental mercury represents the smallest possible volume for mercury-containing secondary waste, although its liquid and volatile nature might pose disposal problems that would not arise with a solid, nonvolatile waste form. Therefore, conversion to HgS or some other highly insoluble mercury salt could be considered. However, it is possible that the elemental mercury recovered in a mercury-remediation process based on chemistry similar to that described herein might be recycled or sold, not discarded.

The thermal removal of mercury from $Hg_{0.32}MoS_2$ and regeneration of MoS_2 was monitored using differential scanning calorimetry (DSC). The DSC of $Hg_{0.32}MoS_2$ displays an endothermic peak at ~350°C upon heating from 25°C to 600°C, Figure 5. The endothermic peak may correspond to the vaporization of elemental mercury (b.p. = 357°C).[39] This observation is consistent with our explanation that mercuric ion, present in $Hg_{0.32}MoS_2$, is first reduced by an internal redox reaction and then vaporized as metallic mercury from the host MoS_2 layers.

The process of activation, absorption, deactivation, and reactivation, which is illustrated in Figure 6, was demonstrated in practice as described below. A single sample of $Li_{1.3}MoS_2$ was treated with 10.0 mM $Hg(NO_3)_2$ (Mo/Hg^{2+} mole ratio = 2.0) in 0.1 M HNO_3, deactivated at 425°C under vacuum, and reactivated with *n*-BuLi through three complete cycles. The conditions (temperature, time, initial

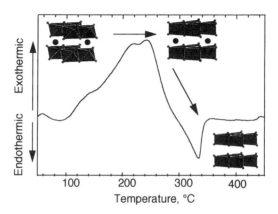

Figure 5   DSC trace and associated phase transitions (1T → 2H) in $Hg_{0.32}MoS_2$ and loss of Hg° in the second, endothermic step.

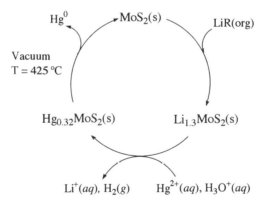

Figure 6   Redox-recyclable cycle for the extraction and recovery of mercury using MoS_2.

molar ratios) were kept constant for all three cycles. Interestingly, more mercury was extracted for the second cycle than for the first, and more was extracted for the third than for the second. This may be due to the increasing value of x measured for the Li_xMoS_2 extractant with successive reactivation steps. We believe that the increase in x is due to faster reductive-intercalation of lithium as the MoS_2 extractant is recycled, which may be attributed to a decrease in the particle size of the recycled material relative to native MoS_2. Scanning electron micrographs revealed the trend in decreasing particle size with each cycle, which is attributable to several cycles of exfoliation and reflocculation that the material experienced during this procedure. A smaller MoS_2 particle size will probably increase the rate of reductive intercalation of MoS_2 by n-BuLi. This in turn would lead to a higher value of x (in Li_xMoS_2) when MoS_2 is treated with n-BuLi for a fixed amount of time. In summary, we have shown that Li_xMoS_2 can be used not only for several complete extraction-deactivation/recovery-reactivation cycles without a significant *loss* of ion-exchange capacity but that the ion-exchange capacity appears to increase. The average stoichiometries $Li_{1.9}MoS_2$ and $Hg_{0.50}MoS_2$ observed for the third cycle of our three-cycle trial resulted in a calculated ion-exchange capacity of 580 mg Hg per gram of $Li_{1.9}MoS_2$ (cf. thioalkylated mesoporous silica which takes up 505 mg Hg per g of material for the first extraction cycle[30a]).

Evidence that confirms the integrity of the host lattice and the activity towards specific guests can be found using XRD. Figure 7 shows the XRD patterns for the extractant precursor, MoS_2, the mercury-included solid, and the deactivated, mercury-free solid. These patterns display the expected changes in lattice spacing and particle size of the host solid for the extraction and deactivation steps. These XRD patterns do not, however, offer any insight into the local structural organization of the extracted contaminant (i.e., the guest ions) within the host lattice. For this, we turned to extended X-ray absorption fine structure (EXAFS) analysis of the materials which were performed at Lawrence Berkeley National Lab.[40]

Shown in Figure 8 are representative Fourier transformed EXAFS data for the extractant precursor, the activated extractant, Li_xMoS_2 solids, and the deactivated extractant. The spectra reveal the coordination environments of the guest species and the host lattice. The mercury guest species are both Hg(II) and Hg(I) ions based on modeling known standards $HgCl_2$, Hg_2Cl_2, and HgS. The geometry of the molybdenum host atoms was transformed from trigonal prismatic to octahedral in the "activated" extractant as well as in the guest-host solids. This organization, however, is metastable and is "reorganized" to a less reducible form of MoS_2 upon heating above 100 °C. From this information, we realized that the best host matrix consists of a MoS_2 lattice (known as the 1-T form) that is "pre-organized" into a particular coordination environment (for Mo) and based on literature precedent, this solid could be activated and deactivated using mild reducing agents (i.e., milder than n-BuLi).

Other MS_2 solids discussed at the beginning of this manuscript were also investigated as possible selective ion-exchange materials: WS_2, TaS_2 and TiS_2. It is well known that the earlier transition metal materials are easily hydrolyzed and so they would probably not be ideal candidates for use with aqueous solutions.[41] For the sake of completeness, their behavior was studied in the presence of Hg(II)-containing aqueous solutions. All three compounds removed substantial amounts of Hg^{2+} from aqueous solution. The results for $Hg_{0.09}WS_2$ indicate that solid $Li_{0.25}WS_2$ is very effective at taking up mercury with the caveat that WS_2 is more difficult to reduce with n-BuLi than MoS_2. If one considers that the maximum amount of $Hg^{2+}(aq)$ that could be taken up by $Li_{0.25}WS_2$ is 0.125 equivalents, then the material is utilizing 75% of its total theoretical capacity. Control experiments showed that unactivated WS_2 extracted only minimal amounts of $Hg^{2+}(aq)$ from an identical waste simulant.

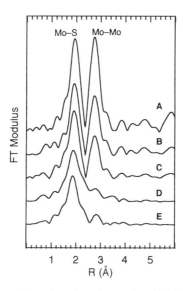

Figure 7   Powder X-ray diffraction data for native (A) MoS_2, (B) the host-guest extractant pair, and (C) the deactivated (heat treated) extractant. Starred peaks are Si or Al.

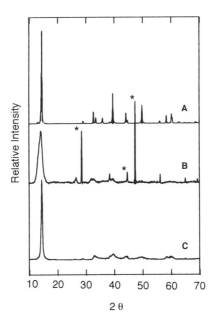

Figure 8   Fourier Transforms of Mo EXAFS data for the solids: (A) MoS_2; (B) $Li_{1.2}MoS_2$ in acid; (C) heat treated $Hg_{0.7}MoS_2$; (D) $Li_{1.2}MoS_2$; (E) $Hg_{0.27}MoS_2$.

The other two activated MS_2 compounds, $Li_{1.5}TiS_2$ and $Li_{1.2}TaS_2$, displayed a very high capacity for $Hg^{2+}(aq)$ with control extraction experiments showing that TiS_2 and TaS_2 removed essentially the same amount of $Hg^{2+}(aq)$ as did their lithium-intercalated analogs. We believe that $Hg^{2+}(aq)$ was not removed by a reversible ion-exchange process, as was the case with MoS_2, but rather by an irreversible hydrolysis of the host lattice. XRD patterns of solids present after $Hg^{2+}(aq)$ was extracted using either $Li_{1.5}TiS_2$ or TiS_2 indicated the presence of only TiS_2 and HgS and an amorphous solid we attribute to a hydrated form of TiO_2.[42] The 001 line for the known phase $Hg_{1.24}TiS_2$ was not observed in the XRD pattern ($2\theta = 10.2°$).[43]

The presence of HgS lines in the XRD patterns can be explained only if S^{2-} anions are formed during the extraction process. Titanium disulfide is known to undergo hydrolysis reactions.[44] One greatly simplified possible reaction is shown in equation (7):

$$TiS_{2(s)} + H_2O_{(l)} \rightarrow TiO^{2+}{}_{(aq)} + 2H^+{}_{(aq)} + 2S^{2-}{}_{(aq)} \qquad (7)$$

If this reaction, or a similar reaction, occurred to an appreciable extent in our experiments, then a significant amount of S^{2-} anions would have been available for the precipitation of HgS ($K_{sp} \sim 10^{-50}$).

High surface area microdimensional nanofibers of MoS_2 have been prepared by us in an effort to probe the viability of using membrane-bound extractants as selective ion-exchange resins for waste remediation.[45] Near-monodisperse microscopic nanostructures of MoS_2 were prepared by thermal decomposition of two different ammonium thiomolybdate molecular precursors, $(NH_4)_2MoS_4$ and $(NH_4)_2Mo_3S_{13}$, within a porous aluminum oxide membrane template. Our low-temperature (450 °C) synthetic route yielded large quantities of near-monodisperse hollow fibers of MoS_2 of uniform size and shape that were ~30 μm long with diameters of up to 300 nm, Figure 9. This template-assisted growth process yielded large quantities of MoS_2 fibers that could be easily isolated from the template (as shown in Figure 9) or held within the template for use later as high surface area extractants imbedded within an aluminum oxide membrane.

Conclusions.

In summary, an entire *repeatable* cycle has been demonstrated: (i) activation of the solid extractant precursor MoS_2 with a chemical reducing agent, n-BuLi, (ii) extraction of Hg^{2+} and other metal ions from aqueous solution, and (iii) recovery of Hg^{2+} from Hg_zMoS_2 and regeneration of MoS_2 by thermal treatment of the mercury-containing solid. This repeatable cycle represents a new paradigm in separation science using redox-recyclable solid-state extractants. We believe that we can generalize this paradigm to include new materials that will outperform currently available methodologies for the extraction and recovery of a particular soft heavy-metal ion from a particular aqueous waste stream. Many existing and new materials remain to be studied, conditions for extraction and recovery to be optimized, and new advances to be made in synthesis to create new composite materials with controllable morphologies and redox/intercalation chemistry.

Acknowledgments.

We would like to thank our coworkers A. E. Gash, L. M. Dysleski, Dr. C. M. Zelenski, and Dr. P. G. Allen, as well as undergraduate students C. J. Flashenriem, A. Kalaveshi, and A. L. Spain, for their dedication to science. Most of this work was supported by a grant from the U.S. DOE (DOE-EMSP DE-FG07-96ER14696). PKD is a 1997 A. P. Sloan Fellow and a Camille Dreyfus Teacher Scholar. Email: pkd@LAMAR.colostate.edu; strauss@chem.colostate.edu.

Figure 9   An SEM image of 300 nm × 30 μm fibers of MoS_2, which have been isolated from the template, shown at 10,000 × magnification (after reference 46).

66

References

1. (a) Tedder, D. W.; Pohland, F. G., Eds. *Emerging Technologies in Hazardous Waste Management III, ACS Symposium Series 518*; American Chemical Society: Washington, DC, 1993, and references therein. (b) Cecille, L.; Casarci, M.; Pietrelli, L., Eds. *New Separation Chemistry Techniques for Radioactive Waste and Other Specific Applications*; Elsevier Applied Science: London, 1991, and references therein.
2. (a) Vandegrift, G. F.; Reed, D. T.; Tasker, I. R., Eds. *Environmental Remediation. Removing Organic and Metal Ion Pollutants, ACS Symp. Series 509*; American Chemical Society: Washington, DC, 1992, and references therein. (b) Clifford, D. A.; Zhang, Z. *Ion. Exch. Technol.* **1995**, 1. (c) Miyake, Y. *Zosui Gijutsu* **1994**, *20*, 57. (d) Förstner, U.; Wittmann, G. T. W. *Metal Pollution in the Aqueous Environment*, Springer-Verlag: Berlin, 1979.
3. (a) Daehne, W. *Galvanotechnik* **1994**, *85*, 2995. (b) Wiaux, J. P.; Nguyen, T. *Galvano-Organo* **1991**, *60*, 587.
4. (a) Thornton, J. D. *Science and Practice of Liquid-Liquid Extraction*, Vols. 1 and 2; Clarendon Press: Oxford, 1992. (b) Baird, M. H. I. *Can. J. Chem. Eng.* **1991**, *69*, 1287. (c) Kertes, A. S.; Marcus, Y. *Solvent Extraction Research*; Wiley-Interscience: New York, 1969.
5. (a) Dasgupta, P. K. *Anal. Chem.* **1992**, *64*, 775A. (b) Peters, R. W.; Shem, L. "Separation of Heavy Metals: Removal from Industrial Wastewaters and Contaminated Soil," In *Emerging Sep. Technol. Met. Fuels, Proc. Symp.*; Lakshmanan, V. I.; Bautista, R. G.; Somasundaran, P., Eds.; Miner. Met. Mater. Soc.: Warrendale, PA, 1993, pp 3-64. (c) Abe, M. *Ion Exch. Solvent Extr.* **1995**, *12*, 381. (d) Streat, M. *Ind. Eng. Chem. Res.* **1995**, *34*, 2841.
6. (a) Hilker, J. *WLB, Wasser, Luft Boden* **1994**, *38*, 22. (b) Shrivastava, A. K.; Rupainwar, D. C. *Chem. Environ. Res.* **1992**, *1*, 235.
7. Kirschner, E. M. *Chem. Eng. News* **1994**, June 20, p 13.
8. Shriver, D. F.; Atkins, P.; Langford, C. H. *Inorganic Chemistry, 2nd Ed.*, Freeman: New York, 1994, pp 212.
9. RCRA = Resource Conservation and Recovery Act; SDWA = Safe Drinking Water Act.
10. Reed, D. T.; Tasker, I. R.; Cunnane, J. C.; Vandegrift, G. F. "Environmental Restoration and Separation Science," in Vandegrift, G. F.; Reed, D. T.; Tasker, I. R., Eds. *Environmental Remediation. Removing Organic and Metal Ion Pollutants, ACS Symposium Series 509*; American Chemical Society: Washington, DC, 1992.
11. (a) Babad, H.; Fulton, J. C.; DeFigh-Price, B. C. *A Strategy for Resolving High-Priority Hanford Site Radioactive Waste Storage Tank Safety Issues*, report WHC-SA-1661-FP, Westinghouse Hanford Company: Richland, WA, 1993. (b) Barker, S. A.; Thornhill, C. K.; Holton, L. K. *Pretreatment Technology Plan*, report WHC-EP-0629, Westinghouse Hanford Company: Richland, WA, 1993. (c) Chidambariah, V.; Travis, C. C.; Trabalka, J. R.; Thomas, J. K. "Risk-Based Prioritization for the Interim Remediation of Inactive Low-Level Liquid Radioactive Waste Underground Storage Tanks at Oak Ridge National Laboratory, Oak Ridge, Tennessee," report ORNL/ER-84, Oak Ridge National Laboratory: Oak Ridge, TN, 1992.
12. (a) Todd, T. Presentation at ESPIP Technical Integration and Exchange Meeting, Gaithersburg, MD, January 1995. (b) *Integrated Data Base for 1992 U.S. Spent Fuel and Radioactive Waste Inventories, Projections, and Characteristics*, report DOE/RW-0006 (Rev. 8), July 1992.
13. Mielke, H. W. *Env. Geochem. Health* **1994**, *16*, 123.

14. Lee, A. Y.; Wethington, A. M.; Miller, V. R. *Proc. Ind. Waste Conf.* **1994**, *48*, 375.
15. Peters, R. W.; Li, W.; Miller, G.; Brewster, M. D.; Patton, T. L.; Martino, L. E. *Hazard. Ind. Wastes* **1995**, *27*, 632.
16. Parrish, C. S.; Uchrin, C. G. *Environ. Toxicol. Chem.* **1990**, *9*, 559.
17. (a) Cheam, Venghout; Lechner, J.; Desrosiers, R.; Sekerka, I. *J. Great Lakes Res.* **1995**, *21*, 384. (b) Miloshova, M. S.; Seleznev, B. L.; Bychkov, E. A. *Sens. Actuators, B* **1994**, *19*, 373, and references therein. (c) Sager, M. *Toxicol. Environ. Chem.* **1994**, *45*, 11.
18. (a) Hosokawa, Y.*Water Sci. Technol.* **1993**, *28*, 339. (b) Saifutdinov, M. M.; Oliger, T. A.; Fomina, Z. S.; Pavlova, S. A. *Gig. Sanit.* **1992**, 19. (c) Baeyens, W. *Trends Analyt. Chem.* **1992**, *11*, 245. (d) Fitzgerald, W. F.; Clarkson, T. W. *Environ. Health Perspect.* **1991**, *96*, 159.
19. Qiang, T.; Xiao-Quan, S.; Jin, Q.; Zhe-Ming, N. *Anal. Chem.* **1994**, *66*, 3562.
20. Wasay, S. A.; Arnfalk, P.; Tokunaga, S. *J. Hazard. Mat.* **1995**, *44*, 93.
21. Tedder, D. W. *Sep. Purif. Methods* **1992**, *21*, 23.
22. Rupprecht, S.; Franklin, S. J.; Raymond, K. N. *Inorg. Chim. Acta* **1995**, *235*, 185.
23. Yordanov, A. T.; Mague, J. T.; Roundhill, D. M. *Inorg. Chem.* **1995**, *34*, 5084.
24. Wang, S.; Elshani, S.; Wai, C. M. *Anal. Chem.* **1995**, *67*, 919.
25. (a) Gandhi, M. N.; Khopkar, S. M. *Anal. Chim. Acta* **1992**, *270*, 87. (b) Tsuda, J.; Sekine, T. *Proc. Symp. Solvent Extr.* **1994**, 53. (c) Mikaelyan, D. A.; Artsruni, V. Z.; Khachatryan, A. G. *J. Anal. Chem.* **1995**, *50*, 149.
26. (a) Buckley, L. P.; Vigayan, S.; McConeghy, G. J.; Maves, S. R.; Martin, J. F. *Removal of Soluble Toxic Metals from Water*, report AECL-10174, Atomic Energy of Canada Limited: Chalk River, Ontario, Canada, 1990. (b) Chaufer, B.; Deratani, A. *Nucl. Chem. Waste Manage.* **1988**, *8*, 175.
27. Misaelides, P.; Godelitsas, A.; Charistos, V.; Ioannou, D.; Charistos, D. *J. Radioanal. Nucl. Chem.* **1994**, *183*, 159.
28. Newton, J. P. "Advanced Chemical Fixation of Organic and Inorganic Content Wastes," In *Hazardous Waste: Detection, Control, Treatment*, Abbou, R., Ed.; Elsevier: Amsterdam, 1988, p 1591.
29. Mercier, L.; Detellier, C. *Env. Sci. Tech.* **1995**, *29*, 1318.
30. (a) Feng X.; Fryxell G.E.; Wang L.-Q.; Kim, A.Y.; Liu, J.; Kemner, K.M. *Science* **1997**, *276*, 923. (b) Liu, J.; Feng, X.; Fryxell, G. E.; Wang, L.-Q.; Kim, A. Y.; Gong, M. *Adv. Mater.* **1998**, in press.
31. (a) Strauss, S. H. *Abstacts of Papers*, 214th National Meeting of the American Chemical Society, Las Vegas, NV; American Chemical Society: Washington DC, 1997; Abstract I&EC060. (b) Clark, J. F.; Clark, D. L.; Whitener, G. D.; Schroeder, N. C.; Strauss, S. H. *Environ. Sci. Technol.* **1996**, *30*, 3124. (c) Chambliss, C. K.; Odom, M. A.; Morales, C. M. L.; Martin, C. R.; Strauss, S. H. *Anal. Chem.* **1998**, *70*, 757. (d) Strauss, S. H. "Redox-Recyclable Extraction and Recovery of Heavy Metal Ions and Radionuclides From Aqueous Media," *Metal Ion Separation and Preconcentration: Progress and Opportunities*; Bond, A. H.; Dietz, M. L.; Rogers, R. D., Eds.; ACS Symposium Series 716; American Chemical Society: Washington, DC, 1998, in press.
32. Hulliger, F. *Structural Chemistry of Layer-Type Phases*; Reidel: Dordrecht, 1976, p 377.
33. See for example (a) Rouxel, J.; Trichet, L.; Chevalier, P.; Colombet, P.; Ghaloun, O. A. *J. Solid State Chem.* **1979**, *29*, 311. (b) Rouxel, J. *J. Solid State Chem.* **1976**, *17*, 223.
34. (a) Gee, M. A.; Frindt, R. F.; Joensen, P.; Morrison, S. R. *Mat. Res. Bull.* **1986**, *21*, 543. (b) Miremadi, B. K.; Morrison, S. R. *J. Appl. Phys.* **1988**, *63*, 4970. (c) Miremadi, B. K.; Morrison, S. R. *J. Appl. Phys.* **1990**, *67*, 1515. (d)

Miremadi, B. K.; Cowan, T.; Morrison, S. R. *J. Appl. Phys.* **1991**, *69*, 6373. (e) Miremandi, B. K., Morrison, S. R. *J. Appl. Phys.* **1990**, *67*, 1515.

35. (a) Whittingham, M. S. *Prog. Solid St. Chem.* **1978**, *12*, 41. (b) Gerischer, H.; Decker, F.; Scrosati, B. *J. Electrochem. Soc.* **1994**, *141*, 2297. (c) Linden, D. *Handbook of Batteries;* 2nd ed.; McGraw Hill: New York, 1995, section 36.10.

36. Hibma, T. in "Structural aspects of Monovalent Cation Intercalates of Layered Dichalcogenides" in *Intercalation Chemistry*, M. S. Whittingham and A. J. Jacobson, Eds., Academic Press: NY, 1982, 285-314.

37. Selected chapters by J. Rouxel, R. Schöllhorn, D. W. Murphy, et al., W. R. McKinnon, and R. F. Frindt in *Nato Advanced Science Institutes Series, Ser. B, Chemical Physics of Intercalation* Legrand, A. P. and Flandois, S., Eds., Plenum Press: NY, 1987.

38. Gash, A. E.; Spain, A. L.; Dysleski, L. M.; Flaschenriem, C. J.; Kalaveshi, A.; Dorhout, P. K.; Strauss, S. H. *Env. Sci. Tech.* **1998**, *32*, 1007.

39. *Handbook of Chemistry and Physics*; Lide, D.R., Ed.; 72nd Ed.; CRC Press: Boca Raton, 1992.

40. (a) Gash, A.E.; Allen, P. G.; Dorhout, P.K.; Strauss, S.H., work in preparation. (b) P. K. Dorhout; S. H. Strauss; Gash, A. E.; Spain, A. S.; Kalaveshi, A. *Abstracts of Papers*, 214th National Meeting of the American Chemical Society, Las Vegas, NV; American Chemical Society: Washington DC, 1997; Abstract I&EC049.

41. (a) Rouxel, J.; Danot, M.; Bichon, J. *Bull. Soc. Chem. Fr.* **1971**, *11*, 3930. (b) Trichet, L.; Jérome, D.; Rouxel, J. *C. R. Acad. Sc. Paris, Ser. C.* **1975**, *280*, 1025. (c) Whangbo, M.-H.; Rouxel, J.; Trichet, L. *Inorg. Chem.* **1985**, *24*, 1824. (d) Ganal, P.; Olberding, W.; Butz, T.; Ouvrard, G. *Solid State Ionics* **1993**, *59*, 313.

42. JCPDF reference numbers: 42-1408 (cinnabar); 06-0261 (metacinnabar).

43. (a) Ganal, P.; Moreau, P.; Ouvrard, G.; Sidorov, M.; McKelvy, M.; Glaunsinger, W. *Chem. Mater.* **1995**, *7*, 1132. (b) Sidorov, M.; McKelvy, M.; Sharma, R.; Ganal, P.; Moreau, P.; Ouvrard, G. *Chem. Mater.* **1995**, *7*, 1140. (c) McKelvy, M.; Sidorov, M.; Marie, A.; Sharma, R.; Glaunsinger, W. *Chem. Mater.* **1994**, *6*, 2233. (d) Ong, E. W.; McKelvy, M. J.; Ouvrard, G.; Glaunsinger, W. S. *Chem. Mater.* **1992**, *4*, 14.

44. Martinez, H.; Autiel, C.; Gonbeau, D.; Loudet, M.; Pfister-Guillouzo, G. *Appl. Surface Sci.* **1996**, *93*, 231.

45. Zelenski, C. M.; Dorhout, P. K. *J. Am. Chem. Soc.* **1998**, *120*, 734.

Chapter 6

New Donors and Acceptors for Molecule-Based Magnetism Research

Brenda J. Korte[1], Roger D. Sommer[1], Scott P. Sellers[1], Mahesh K. Mahanthappa[1], William S. Durfee[2], and Gordon T. Yee[1]

[1]**Department of Chemistry and Biochemistry, University of Colorado, Boulder, CO 80309**
[2]**Department of Chemistry, Buffalo State College, Buffalo, NY 14222**

Candidate building blocks for producing new charge-transfer molecule-based magnetic materials are described. The importance of design criteria such as electrochemical suitability and tunability by organic and inorganic synthetic methods is discussed. A summary of two diagnostic tools available to the magnetochemist is presented. The utility of these techniques is illustrated using examples of several new magnetic charge-transfer salts derived from new donors or acceptors that have recently been synthesized.

At the present time, it is still quite difficult to design and synthesize molecular or molecule-based solids that display three-dimensional ferromagnetic order. Various successful strategies for producing ferrimagnetic, covalently-linked, one-dimensional chain compounds have been reported. However, these materials must also possess less easily designed ferromagnetic coupling between the chains in the remaining two dimensions to exhibit a net moment in the absence of an applied field (*1,2*). Between the chains, relatively weak van der Waals and/or hydrogen bonding furnish the pathway for interaction. The recently discovered ferrimagnetic Prussian-blue analogs have strong coupling in three dimensions and hence high Néel or Curie temperatures, but have lost much of their "molecular" character (*3,4*). The largest organic "molecule" in these compounds is cyanide, although these compounds may still be tuned by varying the component transition metals.

In light of these challenges, we have embarked on a strategy based on the construction of a library of electron donor and acceptor building blocks. Using solution electrochemistry as a guide, these donors and acceptors may be reacted, pair-wise, to produce paramagnetic charge-transfer salts with *reasonably predictable numbers of unpaired electrons*. In the solid state, these unpaired electrons will reside in close proximity to each other (because of Coulomb interactions and possibly coordinate covalent bonds) and will hopefully be poised to couple intermolecularly. Although at present we cannot predict whether the interaction will be ferromagnetic, antiferromagnetic or negligibly small, we can quickly screen compounds using SQUID magnetometry. In principle, each newly discovered donor may be paired with many available acceptors, and vice versa, to produce a new and possibly ferromagnetic solid. Also, many of these new molecules should be tunable using the techniques of the synthetic chemist. Redox potentials, sterics and polarity are some of the characteristics that may be manipulated.

Magnetic Measurements

The usual screening tool for potential molecular magnets is the measurement of magnetization as a function of temperature (M vs. T) utilizing the technique of dc magnetometry. Magnetization (normalized to a per mole basis) is related to molar susceptibility by the equation $M/H = \chi_M$ where H is the applied magnetic field in gauss (typically 100 to 5000 G). For ideal paramagnets, the plot of $\chi_M T$ vs. T derived from these data should be linear with zero slope because they obey the Curie Law ($\chi_M = C/T$ where $C = g^2 S(S+1)/8$). The value of χT at room temperature can often be used to infer the number of unpaired electrons, S, and the g value. Two criteria must considered in assigning g and S: typically the g value is close to 2.0, and the number of unpaired electrons per formula unit must be a whole number that makes good chemical sense.

The presence of ferromagnetic interactions between unpaired spins on the molecules may be indicated by deviations from linearity at low temperatures to higher values of χT. Because the coupling in molecular and molecule-based compounds is typically weak, deviations from zero slope at high temperatures should be small, i. e. the compounds should behave as paramagnets in this temperature region. When interactions between unpaired spins on the molecules are present, a plot of χ^{-1} vs. T will generally produce a straight line with non-zero x-intercept according to the Curie-Weiss law ($\chi_M = C/(T - \theta)$), where the value of θ is a measure of the degree of interaction. When the interactions are ferromagnetic in nature, the value of θ should be positive.

Although the increase in χT at low temperatures and the observation of a positive θ are indicators of ferromagnetic interactions, they are insufficient to characterize a compound as a ferromagnet, because ferromagnetism results from a three-dimensional phase transition. To observe and characterize this transition, the ac susceptibility is measured. In this technique, an oscillating magnetic field (typically 5 G amplitude at frequencies ranging from 1 Hz to 1000 Hz) is applied to the sample, and the susceptibility in-phase (real) and 90° out-of-phase (imaginary) with the drive field is monitored. The phase transition to the ferromagnetic state is marked by a peak in the in-phase component accompanied by the appearance of a non-zero out-of-phase signal (*5*). These signals should be frequency independent for a true ferromagnet.

Charge-transfer Salts

Charge-transfer compounds, which result from the reaction of two neutral building blocks, were first investigated for their interesting electrical conductivity properties. Compounds such as tetrathiafulvalene-tetracyanoquinodimethane (TTF-TCNQ), in which electrons are delocalized in π stacks, ushered in an era of intense investigation. This early period culminated in the discovery of superconductivity in 1980 in the compound derived from the electrochemical oxidation of tetramethyltetraselenafulvalene in the presence of hexafluorophosphate to give $(TMTSeF)_2PF_6$ (*6*).

TTF "D"   TCNQ "A"

Guidelines for designing conductors center on the necessity of non-integral charge transfer, ρ, where typically $0 \leq \rho \leq 1$ and the requirement for segregated stacks. Degree of charge transfer is a measure of the final charge on the donor (+ρ) or acceptor (-ρ) assuming a 1:1 complex. Electrically conductive charge-transfer solids are most often found as segregated stacks which are characterized by structures in which donors and acceptors crystallize in separate columns, as shown below.

D ——— A ———	D ——— A ———
D ——— A ———	A ——— D ———
D ——— A ———	D ——— A ———
D ——— A ———	A ——— D ———
D ——— A ———	D ——— A ———
segregated stacks	mixed stacks

These concepts were further refined to provide useful, empirical, criteria based on the difference in solution redox potential between the donor and acceptor. The quantity $\Delta E^\circ_{a-d} (= E^\circ_a - E^\circ_d)$ is a measure of the propensity for an electron to be transferred from donor to acceptor. It has been shown that generally(7)

1) good donors paired with good acceptors give rise to ionic, non-conductive solids, where $\rho = 1$ and ΔE°_{a-d} is large and positive
2) weak donors paired with weak acceptors yield neutral non-conductors, where $\rho = 0$ and ΔE°_{a-d} is large and negative
3) moderate donors paired with moderate acceptors give metallic conductors, $0 < \rho < 1$ and ΔE°_{a-d} is near zero.

In contrast to the recipe for producing conductivity (criterion 3), magnetically interesting charge-transfer salts are likely to be found in those combinations that satisfy criterion 1 and which crystallize in mixed stacks. For example, if the neutral donor and acceptor are each diamagnetic before the electron transfer, their reaction in a relatively non-polar solvent will produce ionic radical cation-radical anion pairs which will crystallize in mixed stacks because of favorable Madelung energy. Assuming no further charge transfer is allowed, the ionic product maximizes the number of effective unpaired spins per formula unit. This condition is important, however, because additional charge transfer (i.e. a donor that contributes two electrons to an acceptor producing doubly-charged species) could give rise to a diamagnetic ion (e.g. if the acceptor LUMO is non-degenerate).

$$D + A \longrightarrow D^{\bullet/+} + A^{\bullet/-} \xrightarrow{\mathllap{X}} D^{2+} + A^{2-}$$

Note that because of the presence of d orbitals, building blocks containing transition metals may possess more than one unpaired electron so that it is not necessary for the neutral building blocks to be diamagnetic.

The first ferromagnetic charge-transfer (CT) salt to display hysteresis was prepared in this manner. Decamethylferrocenium tetracyanoethenide, [Fe(Cp*)$_2$][TCNE], reported by Miller and coworkers in 1987, was prepared by the simple addition of solutions of Fe(Cp*)$_2$ and TCNE in acetonitrile (8). One unpaired electron is associated with each anion and cation. Below 4.2 K, these spins align ferromagnetically to give a three-dimensionally ordered solid. Prior to this discovery, the analogous CT salt based on Fe(Cp*)$_2$ and the acceptor, TCNQ, was reported to be a metamagnet (9). Later, Fe(Cp*)$_2$ and hexacyanobutadiene (HCBD) were found also to react to form a ferromagnetic phase (10).

A breakthrough occurred in 1990 when Hoffman coworkers reported that a salt built from decamethylmanganocene, [Mn(Cp*)$_2$][TCNQ], could be synthesized by a metathetical route involving the preformed cation, $Mn(Cp^*)_2^+$, and $TCNQ^{\cdot/-}$ radical anion with a reported T_c of 6.2 K (11). This development added to the synthetic toolkit and to the palette of donors which give rise to ferromagnetic solids. Shortly thereafter, $[Mn(Cp^*)_2][TCNE]$ (12) and $Cr(Cp^*)_2$ analogues of these compounds were described (13,14,15). Table I lists the Curie temperatures of all these CT salts (16).

Table I. Curie temperatures for some charge-transfer salts

$[Cr(Cp^*)_2][TCNE]$	4.0 K		$[Cr(Cp^*)_2][TCNQ]$	3.1 K
$[Mn(Cp^*)_2][TCNE]$	8.8 K		$[Mn(Cp^*)_2][TCNQ]$	6.2 K
$[Fe(Cp^*)_2][TCNE]$	4.2 K		$[Fe(Cp^*)_2][TCNQ]$	3.0 K

A second structural type of donor-acceptor magnet is one in which the acceptor binds to the coordinatively-unsaturated donor either before or after charge transfer. An example was discovered by Basolo and coworkers who originally prepared the linear chain complex, meso-(tetraphenylporphyrinato)manganese (III) tetracyanoethenide), $(MnTTP)\mu$-TCNE, and showed it to possess a positive θ (17). Miller and coworkers subsequently demonstrated that $(MnTTP)\mu$-TCNE orders ferrimagnetically within the chain and ferromagnetically between chains below 18 K (18). This compound may be thought of as a donor-acceptor compound because its synthesis involves transfer of an electron from a neutral Mn(II)TPP molecule to a neutral TCNE molecule. If the electron transfer is outer-sphere, then this redox step is followed by coordination of $TCNE^{\cdot/-}$ to a cationic Mn(III) ($S = 2$) center. The net effect of this sequence of events is the synthesis of a paramagnetic 1-D coordination polymer in which spin resides on the metal and bridging TCNE. As with the other 1-D chains, weak interchain coupling may give rise to bulk ordering. Related solids in which Miller has varied the porphyrin macrocycle are providing a family of compounds, many of which show bulk ferromagnetic order (19,20). The use of redox potentials for predicting reactivity is somewhat less useful in this class of reactions because oxidation of the metal center may depend strongly on the nature of the coordinating ligand. However, MnTPP is certainly quite reducing; its Mn(II/III) couple occurs at -0.27 V vs. SCE.

Despite these successes, the donor-acceptor strategy has been slowed by a lack of new examples of donors and acceptors to use as raw materials. Figure 1 illustrates most of the building blocks that have been used successfully to produce molecular ferromagnetic phases along with a few which have produced metamagnets and ferrimagnets. Note that almost all of these are commercially available compounds. The identification and synthesis of new donor families beyond decamethylmetallocenes and new acceptors families beyond cyanocarbons represent research opportunities. Along these lines, Kahn and coworkers have recently reported the acceptor nickel(III) bis(2-oxo-1,3-dithiole-4,5-dithiolate) which, when paired with decamethylferrocene, shows ferromagnetic interactions, though no phase transition (21).

Results and Discussion

One Electron Donors. The search for new donors, is greatly aided by the use of electrochemistry as a guide. Strong donors with large negative (+1/0) couples are desirable because they can be paired with the widest range of acceptors and still satisfy criterion 3. Of course, like decamethylmanganocene, this often means that strong donors will be air-sensitive. Other attributes of the metallocenes that should be mimicked include reversible electrochemistry, unpaired electrons in the monocationic state, a π electron systems for mediating communication between spins and tunability via chemical modification.

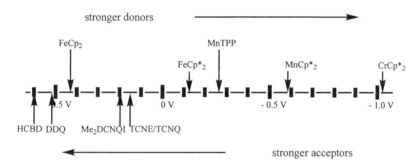

Figure 1. Electrochemical potentials for Donor (+1/0) and Acceptor (0/-1) couples vs. SCE.

Ni(SQDI)$_2$. We are investigating a series of compounds based on the parent complex bis(semiquinonediimine)nickel(II), Ni(SQDI)$_2$. The tunability of this system comes from the ease with which peripheral substitutions to the benzene ring (R) may be made. Each monoanionic ligand is derived from a 4,5-substituted *ortho*-phenylenediamine and exists as a formal radical anion possessing one unpaired electron. Strong antiferromagnetic coupling of these spins mediated by the square planar d^8, diamagnetic, nickel(II) cation, results in an overall diamagnetic complex.

Some of these donors were first prepared by Feigl and Fürth (*22*) and studied electrochemically by Balch and Holm (*23*). The cyclic voltammetry presented here is consistent with these previously reported polarigraphic experiments. Miles and Wilson previously studied these complexes as donors in reactions with TCNQ and other acceptors, hoping to synthesize conducting solids. However, only semiconducting behavior was observed, presumably because mixed stacks were obtained (*24*). The +1/0 couples vs. SCE for several members of this family of compounds are given in Table II. The site of oxidation is generally believed to possess mostly ligand character as evidenced by the rather small shift in E° which results from substituting Pd(II) for Ni(II) and from EPR studies of the radical cation (*23*).

Table II. Nickel(III/II) couples vs. SCE for square planar semiquinonediimine complexes

Ni(SQDI)$_2$	0.069 V
Ni(EtSQDI)$_2$	-0.03 V
Ni(Me$_2$SQDI)$_2$	-0.11 V
Ni(Et$_2$SQDI)$_2$	-0.159 V
Ni((MeO)$_2$SQDI)$_2$	-0.262
Pd(SQDI)$_2$	0.011

Based on the position of the +1/0 couple (Figure 1), the reaction of Ni(SQDI)$_2$ would be expected to yield incomplete charge transfer to TCNE. Reaction of equimolar solutions of Ni(SQDI)$_2$ and TCNE in THF at room temperature gives a solid whose magnetic properties indicate fewer than two unpaired electrons per formula unit. However, this result is complicated by the fact that the unsubstituted complex is rather insoluble. This could inhibit complete reaction and masquerade as insufficient driving force for complete charge transfer.

Addition of four methyl groups to the periphery produces Ni(Me$_2$SQDI)$_2$ and results in a shift of the +1/0 couple to more negative potentials. Reaction of this derivative with TCNE, again in THF, gives rise to the expected room temperature χT product and, more importantly, deviations of the data from linearity, indicating the presence of ferromagnetic interactions between the spins (Figure 2). The ac susceptibility measurements on [Ni(Me$_2$SQDI)$_2$][TCNE] serve to confirm the presence of a phase transition (Figure 3). The peak in χ' at 10 K is indicative of a phase transition while the appearance of a non-zero χ'' indicates the ferromagnetic nature of this transition. The frequency dependence of these data (Figure 3) may indicate that the solid is not a true ferromagnet but rather a spin-glass (*25*). This result requires further investigation.

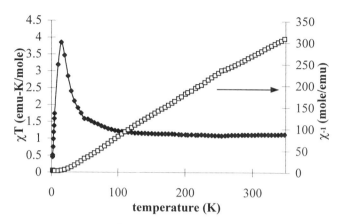

Figure 2. χT vs. T ($\blacklozenge$) and χ^{-1} vs. T ($\square$) for [Ni(Me$_2$SQDI)$_2$][TCNE] measured in 1000 G.

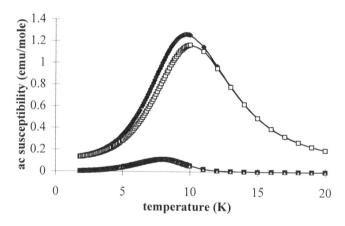

Figure 3. ac Susceptibility data for [Ni(Me$_2$SQDI)$_2$][TCNE] a) χ_{real} 10 Hz ($\blacklozenge$) b) $\chi_{imaginary}$ 10 Hz ($\triangle$) c) χ_{real} 100 Hz ($\square$) d) $\chi_{imaginary}$ 100 Hz ($\blacksquare$).

[Ni(Me$_2$SQDI)$_2$][TCNE] is, unfortunately, very insoluble and the diimine ligand is somewhat labile, so this has hampered our ability to characterize the complex structurally. The IR spectrum shows CN stretches at 2196 and 2167 cm^{-1} which seem to indicate a bridging radical anion as may be seen from data in Table III. This structural motif has been seen previously in Mn(TPP)μ-TCNE and inferred for Cr(TPP)μ-TCNE, as described below. This result was unexpected.

However, if this structure is correct for [Ni(Me$_2$SQDI)$_2$][TCNE], then a significant electronic difference exists between this compound and the aforementioned porphyrinic compounds. In the latter compounds antiferromagnetic coupling is present, whereas in the new nickel compound ferromagnetic interactions are dominant. This result may be rationalized by considering the Goodenough-Kanamori rules and the symmetry of the orbitals on the metal which contain the spin and comparing these to the partially occupied orbital on TCNE$^{•/-}$ assuming octahedral geometry (*26*). For Cr(III) and Mn(III), the partially occupied orbitals (d$_{xz}$, d$_{yz}$) are of π symmetry which overlap with the π* orbital on TCNE$^{•/-}$ which is partially occupied, producing antiferromagnetic coupling. In contrast, for Ni(III), the unpaired electron is in an d$_{z^2}$ or σ type orbital. Unpaired electrons in orbitals of different symmetry (i.e. orbitals that are orthogonal) couple ferromagnetically.

Table III. ν$_{C≡N}$ for TCNE species (in cm^{-1})

TCNE	2260	2225	
TCNE$^{•/-}$ (uncoordinated)	2184	2143	[Fe(Cp*)$_2$][TCNE]
TCNE$^{•/-}$ (bridging)	2192	2147	Mn(TPP)μ-TCNE

CrTPP. As discussed earlier, Miller et al. have found Mn(TPP)μ-TCNE to undergo a phase transition to a ferromagnetically ordered phase below 18 K and have also investigated complexes in which tetraphenylporphyrin is replaced by other porphyrins (*19, 27*).

One other way that this system can be chemically modified is by changing the central metal and hence the number of unpaired electrons present. Thus, we prepared Cr(TPP)μ-TCNE by substituting four coordinate *meso*-tetraphenyl-porphyrinatochromium(II), Cr(II)TPP, for five coordinate Mn(TPP)(py) in the published synthetic procedure for Mn(TPP)μ-TCNE (*18*). Cr(II)TPP is obtained by the reversible reduction of chloro-*meso*-tetraphenylporphyrinato-chromium(III), Cr(III)TPPCl, with bis(2,4-pentanedionato) chromium(II) (Cr(acac)$_2$) (*28*). Based on the infrared stretching frequencies in the nitrile region (2169 cm^{-1}, 2214 cm^{-1}), the TCNE radical anions in Cr(TPP)μ-TCNE are N-bound and σ-coordinated as in Mn(TPP)μ-TCNE (2147 cm^{-1}, 2192 cm^{-1}). The two compounds are assumed to be isostructural although we have been unable to obtain single crystals suitable for structure analysis.

A plot of the product of molar susceptibility and temperature (χT) for a polycrystalline sample of Cr(TPP)μ-TCNE is shown in Figure 4. The room temperature value of χT (2.09 emu-K/mol) is consistent with non-interacting S=3/2 (Cr(III)) and S=1/2 (TCNE radical anion) spins. The expected value for such a system is 2.25 emu-K/mol, assuming isotropic g=2 for each spin center. This is reasonable if the ground state of Cr(III) is ^{4}A$_{1g}$. As the temperature is decreased, χT passes through a minimum at 250 K (characteristic of ferrimagnetic chain compounds) and then diverges to a value of 4.88 emu-K/mol at 10 K. This is followed by a decrease in χT as the temperature is lowered further. A plot of inverse susceptibility (χ^{-1}) versus temperature shows Curie-Weiss behavior in two distinct temperature regions. A fit to the high temperature data (250 K< T < 350 K) results in θ = -106 K whereas the low temperature data (100 K< T < 245 K) gives a θ of +26 K. These values may be compared to those of the manganese analog

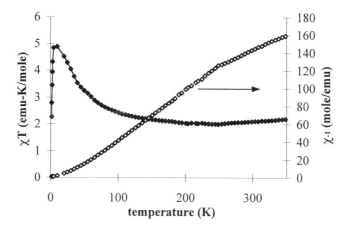

Figure 4. χT vs. T ($\blacklozenge$) and χ^{-1} vs. T ($\square$) for Cr(TPP)μ-TCNE measured in 1000 G.

(which also shows two regimes) where data fit above 280 K give a θ of -15 K and data fit between 115 K and 250 K gives a θ of +61 K.

In a mean field model, T_c is predicted to be proportional to $[2S(S+1)zJ]/3k$ where z is the number of nearest neighbors and J is a measure of the coupling. Since the above model is only valid for a single-spin system, the high temperature (intrachain) data must be interpreted using an effective spin which is assumed to be larger in the Mn analogue. We also assume that within the chain, the number of nearest neighbors (two) is the same for both compounds. Then, from the much more negative high temperature θ for Cr(TPP)μ-TCNE, we can conclude that the antiferromagnetic intrachain exchange constant, J_{intra}, must be significantly more negative for the chromium compound than for the manganese compound.

At low temperatures, where the antiferromagnetic interaction between spins centers is large, the effective spin system can be more simply described by $S = 1$ and $S = 3/2$, respectively, for each Cr(III)•TCNE and Mn(III)•TCNE formula unit. The roughly factor of two difference in the observed low temperature θ's may then be neatly rationalized by the difference in the $S(S+1)$ term, suggesting that the interchain coupling, J_{inter}, is comparable for both systems.

The field dependence of the magnetization, M, was measured at 4 K up to a field of 5 T (Figure 5). M rises quickly up to 1 T and then begins to saturate, reaching a value of 9228 emu-G/mol at 5 T. This value is approaching the saturation magnetization for isotropic $S = 1$ ($M_{sat} = 11170$ emu-G/mol with g = 2), indicating that there is antiferromagnetic coupling between the Cr(III) and TCNE$^{•/-}$ spin centers. The magnetization rises more quickly than expected for an $S = 1$ spin system as calculated from the Brillouin function, suggesting some intermolecular coupling.

The results of the ac susceptibility measurement are shown in Figure 6. The real component of the susceptibility (χ') increases with decreasing temperature, perhaps beginning to saturate approaching 1.7 K, however χ'' is always zero. This indicates that Cr(TPP)μ-TCNE, unlike its Mn analog, does not order above 1.7 K.

Unfortunately, attempts to grow single crystals by methods similar to those reported for the Mn compound (involving dissolution of the polymer into a non-coordinating solvent system and cooling) have, so far, resulted in a fine black powder unsuitable for diffraction. Dissolution of this material requires disassembling the polymer via retro-charge transfer reforming the initial chromium (II) and neutral TCNE precursors. The greater stability of octahedral Cr(III) over octahedral Mn(III) and the much more negative reduction potential (vs. SCE) of Cr(III) [-0.27 V compared to -0.86 V, for Mn(III) and Cr(III)[18] respectively] severely inhibits this mechanism. Slow diffusion of toluene solutions of the two precursors together in a glass tube at -15 °C or through a porous frit, both result in the formation of a very fine black powder. Attempts to dissolve the polymer in coordinating solvents yield powdery Cr(III) products.

The results for this compound are admittedly ambiguous, in part, because its magnetic properties seem to be interesting at temperatures below the capabilities of the SQUID and because we have been unable to get more thorough structural information. Pairing CrTPP with a poorer acceptor (such as DDF, below) may allow us to crystallize a product if indeed, the extreme ionic nature of the metal-bridging ligand bond in Cr(TPP)μ-TCNE is what makes this compound intractable.

One Electron Acceptors. When considering the design and synthesis of new acceptors, one can observe that the great majority of ferromagnetic CT salts are based on tetracyanoethylene rather than on the large family of quinone-based acceptors. The reasons behind this are unclear. One reason may be that quinone-derived radical anions have a propensity to dimerize to yield diamagnetic dianions (*29*).

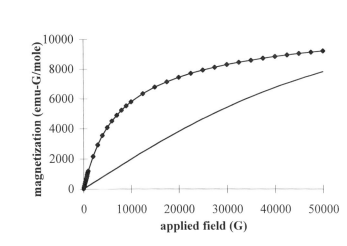

Figure 5. Magnetization vs. applied field for Cr(TPP)μ-TCNE at 4.1 K (♦) and comparison to the Brillouin function (–) calculated for $S = 1$.

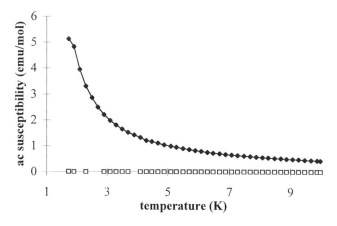

Figure 6. ac Susceptibility data for Cr(TPP)μ-TCNE at 100 Hz. χ_{real} (♦) $\chi_{imaginary}$ (□).

$$2 \text{ TCNQ}^{\cdot/-} \rightarrow [\text{TCNQ}_2]^{2-}$$

From the organic chemist's standpoint, there are considerably more opportunities to modify a quinone than tetracyanoethylene (*30*) In addition to the aforementioned TCNQ and DDQ, possible quinone derivatives include those created by substitution of one or more ring hydrogen atoms with halogen atoms or alkyl groups, and the quinone oxygen atoms by cyanoimine groups (N-C≡ N) to form, for instance, dimethyldicyanoquinodiimine (Me$_2$DCNQI).

DDQ   TCNQ-F$_4$ (Me$_2$DCNQI)

Analogous modifications to tetracyanoethylene are obviously not possible, although hexacyanobutadiene is one related acceptor that has seen limited use. However, it should be possible to replace one or more nitrile groups with other electron-withdrawing groups. A fruitful place to look for ideas is in the Diels-Alder literature which is rife with electron-poor olefins. As always, though, new acceptor candidates should exhibit reversible electrochemistry. Examples, which illustrate failed substitutions, include 1,2-dichloro-1,2-dicyanoethylene and 2,3-bis(trifluoromethyl)fumaronitrile (*31*), both of which probably lose halide upon reduction by one electron.

DDF. An electron deficient olefin which is known to undergo reversible electrochemical reduction is diethyl-2,3-dicyanofumarate (DDF) (*32*). This compound may be synthesized in one step from commercially available ethylcyanoacetate by an oxidative dimerization and may be accomplished by reaction with either SOCl$_2$ or KI/Al$_2$O$_3$ (*33,34*)

The first reduction occurs at -0.240 V vs. SCE which is approximately 0.4 V more negative than TCNE and underscores the almost irreplaceable electron withdrawing ability of the nitrile group. The relatively unfavorable position of this first reduction for DDF necessitates its pairing with a good electron donor such as Mn(Cp*)$_2$. However, this illustrates a strength of the donor acceptor strategy: weak acceptors are managed by matching them with strong donors. There is much room to explore in "potential space" for new acceptors as may be seen in Figure 1.

Reaction of equimolar solutions of Mn(Cp*)$_2$ and DDF in hexane at room temperature results in an immediate color change to brown (similar to the analogous reaction with TCNE) and precipitation of the product. The plot of χT vs. T is shown in Figure 7. It indicates a well-behaved charge transfer salt and a room temperature χT product consistent with other previously reported salts of

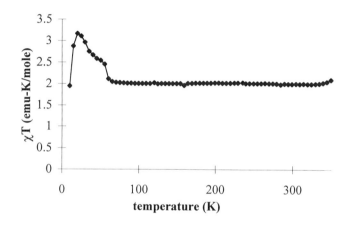

Figure 7. χT vs. T for [Mn(Cp*)₂][DDF] measured in 5000 G.

$Mn(Cp^*)_2^+$ including those with TCNE (*35*), DDQ (*36*) and dimethyl-*p*-dicyanoquinodiimine (Me_2DCNQI) (*37*). In all of these compounds three similar features are observed: a) the donor possesses two unpaired electrons and the acceptor, one, b) the g value for the donor deviates somewhat above the free electron value indicating incomplete quenching of orbital angular momentum and c) the compounds show ferromagnetic coupling of the unpaired spins. In the plot shown for $[Mn(Cp^*)_2][DDF]$, there are hints of ferromagnetic interactions at low temperatures which are the subject of current research. In particular the low temperature synthesis of this compound must be explored given the substitutional lability of the Cp* rings in $Mn(Cp^*)_2$.

Other acceptors we are investigating include tetramethyl- and tetraethyl-ethylenetetracarboxylate (*38*) , dicyanofumaric acid and 3,4-dicyano-3-hexene-2,5-dione.

Conclusions

The donor-acceptor approach has allowed us to prepare a number of new magnetic compounds. Some of the compounds show evidence for ferromagnetic coupling and ordering. More importantly, the ability to make chemical modifications suggests that many more building blocks should be available. The compounds described here may be viewed as promising leads for future efforts.

Acknowledgments

We thank Professor R. Huisgen for the gift of 2,3-bis(trifluoromethyl) fumaronitrile. We also thank Dr. Ron Goldfarb for many helpful discussions and the National Institute of Standards and Technology for use of the SQUID magnetometers. We acknowledge the Donors of the Petroleum Fund, administered by the American Chemical Society, for support of this research.

Literature Cited

1. Nakatani, K.; Bergerat, P.; Codjovi, E.; Mathoniere, C.; Pei, Y. and Kahn, O. *Inorg. Chem.* **1991**, *30*, 3977-3978.

2. Caneschi, A.; Gatteschi, D.; Renard, J. P.; Rey, P. and Sessoli, R. *Inorg. Chem.* **1989**, *28*, 3314.

3. Mallah, T.; Ferlay, S.; Auberger, C.; Helary, C.; L'Hermite, F.; Ouahes, R.; Vaissermann, J.; Verdaguer, M. and Veillet, P. *Mol. Cryst. Liq. Cryst.* **1995**, *273*, 141-151.

4. Entley, W. R.; Treadway, C. R. and Girolami, G. S. *Mol. Cryst. Liq. Cryst.* **1995**, *273*, 153-166.

5. For a more complete description of this technique, see O'Conner, C. J. in *Molecule-Based Magnetic Materials, Theory, Techniques and Applications*, Turnbull, M. M.; Sugimoto, T. and Thompson, L. K., Eds. ACS Symp. Ser. 644, American Chemical Society: Washington, D. C., 1995; pp 44-66.

6. Jerome, D.; Mazaud, A.; Ribault, M. and Bechgaard, K. *J. Phys. Lett,* **1980**, *41*, 95.

7. Ward, M. D. in Electroanalytical Chemistry, Bard, A. J. Ed. Dekker: New York, 1988, 16, 181-312.

8. Miller, J. S.; Calabrese, J. C.; Rommelmann, H.; Chittipeddi, S. R.; Zhang, J. H.; Reiff, W. M. and Epstein, A. J. *J. Am. Chem. Soc.* **1987**, *109*, 769-781.
9. Candela, G. A.; Swarzendruber, L.; Miller, J. S. and Rice, M. J. *J. Am. Chem. Soc.* **1979**, *101*, 2755-2756.
10. Miller, J. S.; Zhang, J. H. and Reiff, W. M. *J. Am. Chem. Soc.* **1987**, *109*, 4584-4592.
11. Broderick, W. E.; Thompson, J. A.; Day, E. P.; Hoffman, B. M. *Science*, **1990**, 249, 401.
12. Yee, G. T.; Manriquez, J. M.; Dixon, D. A.; McLean, R. S.; Grodski, D. M.; Flippen, R. B.; Narayan, K. S.; Epstein, A. J. and Miller, J. S. *Adv. Mater.* **1991**, *3*, 309-311.
13. Broderick, W. E.; Hoffman, B. M. *J. Am. Chem. Soc.* **1991**, *113*, 6334.
14. Miller, J. S.; McLean, R. S.; Vazquez, C.; Calabrese, J. C.; Zuo, F. and Epstein, A. J. *J. Mater. Chem.* **1993**, *3*, 215.
15. Eichhorn, D. M.; Skee, D. C.; Broderick, W. E. and Hoffman, B. M. *Inorg. Chem.* **1993**, *32*, 491.
16. Broderick, W. E.; Liu, X.; Owens, S.; Toscano, P. M.; Eichhorn, D. M. and Hoffman, B. M. *Mol. Cryst. Liq. Cryst.* **1995**, *273*, 17-20.
17. Summerville, D. A.; Cape, T. W. ; Johnson, E. D. and Basolo, F. *Inorg. Chem.* **1978**, *11*, 3297-3300.
18. Miller, J. S.; Calabrese, J. C.; McLean, R. S. and Epstein, A. J. *Adv. Mater.* **1992**, *4*, 498-501.
19. Böhm, A.; Vazquez, C.; McLean, R. S.; Calabrese, J. C.; Kalm, S. E.; Manson, J. L.; Epstein, A. J. and Miller, J. S. *Inorg. Chem.* **1996**, *35*, 3083-3088.
20. Miller, J. S.; Vazquez, C.; Jones, N. L.; McLean, R. S. and Epstein, A. J. *J. Mater. Chem.* **1995**, *5*, 707-711.
21. Ferrouhi, M.; Ouahab, L.; Codjovi, E. and Kahn, O. *Mol. Cryst. Liq. Cryst.* **1995**, *273*, 29-33.
22. Feigl, F. and Fürth, M. *Monatsh. Chem.* **1927**, *48*, 445.
23. Balch, A. L. and Holm, R. H. *J. Am. Chem. Soc.* **1966**, *88*, 5201-5209.
24. Miles, M. G. and Wilson, J. D. *Inorg. Chem.* **1975**, *14*, 2357-2360.
25. Mydosh, J. A. *Spin Glasses, An Experimental Introduction* Taylor and Francis: London, 1993.
26. Kahn, O. *Molecular Magnetism* VCH:New York, 1993 pp 190-191.
27. Miller, J. S. Vazquez, C. and Epstein, A. J. *J. Mater. Chem.* **1995**, *5*, 707.
28. Cheung, S. K.; Grimes, C. J.; Wong, J. and Reed, C. A. *J. Am. Chem. Soc.* **1976**, *98*, 5028-5029.
29. O'Hare, D. and Miller, J. S. *Mol. Cryst. Liq. Cryst.* **1989**, *176*, 381-390.
30. Martin, N.; Segura, J. L. and Seoane, C. *J. Mater. Chem.* **1997**, *7*, 1661-1676.
31. Proskow, S.; Simmons, H. E. and Cairns, T. L. *J. Am. Chem. Soc.* **1966**, *88*, 5254.
32. Mulvaney, J. E.; Cramer, R. J. and Hall, H. K. Jr. *J. Polym. Sci. Polym. Chem.* Vol. 21, 1983 309-314.
33. Ireland, C. J.; Jones, K.; Pizey, J. S. and Johnson, S. *Synth. Commun.* **1976**, *6*, 185-191.
34. Villemin, D. and Ben Alloum, A. *Synth. Commun.* **1992**, *22*, 3169-3179.
35. Yee, G. T.; Manriquez, J. M.; Dixon, D. A.; McLean, R. S.; Grodski, D. M.; Flippen, R. B.; Narayan, K. S.; Epstein, A. J. and Miller, J. S. *Adv. Mater.* **1991**, *3*, 309-311.
36. Miller, J. S.; McLean, R. S.; Vazquez, C.; Yee, G. T.; Narayan, K. S. and Epstein, A. J. *J. Mater. Chem.* **1991**, *1*, 479-480.
37. Miller, J. S.; Vazquez, C.; McLean, R. S.; Reiff, W. M.; Aumüller, A. and Hünig, S. *Adv. Mater.* **1993**, *5*, 448-450.
38. *Org. Synth. Coll.* Vol. IV 1963, 273-275.

PREPARATION AND CHARACTERIZATION OF THIN FILMS

Chapter 7

Chemical Vapor Deposition of Refractory Ternary Nitrides for Advanced Diffusion Barriers

Jonathan S. Custer, Paul Martin Smith, James G. Fleming, and Elizabeth Roherty-Osmun

Microelectronics Development Laboratory, Sandia National Laboratories, Albuquerque, NM 87185

Refractory ternary nitride films for diffusion barriers in microelectronics have been grown using chemical vapor deposition. Thin films of titanium-silicon-nitride, tungsten-boron-nitride, and tungsten-silicon-nitride of various compositions have been deposited on 150 mm Si wafers. The microstructures of the films are either fully amorphous for the tungsten based films, or nanocrystalline TiN in an amorphous matrix for titanium-silicon-nitride. All films exhibit step coverages suitable for use in future microelectronics generations. Selected films have been tested as diffusion barriers between copper and silicon, and generally perform extremely well. These films are promising candidates for advanced diffusion barriers for microelectronics applications.

The manufacturing of silicon wafers into integrated circuits uses many different process and materials. The manufacturing process is usually divided into two parts: the front end of line (FEOL) and the back end of line (BEOL). In the FEOL the individual transistors that are the heart of an integrated circuit are made on the silicon wafer. The responsibility of the BEOL is to wire all the transistors together to make a complete circuit. The transistors are fabricated in the silicon itself. The wiring is made out of metal, currently aluminum and tungsten, insulated by silicon dioxide, see Figure 1. Unfortunately, silicon will diffuse into aluminum, causing aluminum spiking of junctions, killing transistors. Similarly, during chemical vapor deposition (CVD) of tungsten from WF_6, the reactivity of the fluorine can cause "wormholes" in the silicon, also destroying transistors. The solution to these problems is a so-called diffusion barrier, which will allow current to pass from the transistors to the wiring, but will prevent reactions between silicon and the metal.

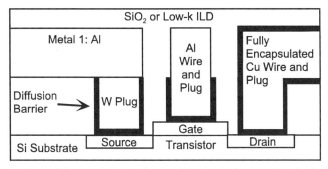

Figure 1. General integrated circuit metallization scheme. On the left is the current industry standard of tungsten vias and aluminum alloy wires. The gate contact shows the potential replacement of the tungsten with aluminum. On the right is the future with Cu completely encapsulated by a diffusion

There are a number of requirements for a diffusion barrier for silicon microelectronics, which have often been reviewed in the literature (*1-6*). As device dimensions shrink, the demands on diffusion barrier performance and processing reliability increase. One set of demands is on the composition and microstructure of the barrier material. The barrier must keep the silicon and metal apart. First, the barrier must remain macroscopically intact as a layer between the silicon and metal layers. This can be accomplished in several ways. A variety of "reactive" barriers have been investigated which will react with the silicon or the metal, although the reaction is self-limiting and part of the barrier will remain intact for the lifetime of the device. As devices get smaller, the barrier must shrink proportionately, so it is more difficult to find materials that continue to work. Second, the barrier can be made of a material that remains in thermodynamic equilibrium with both silicon and the metal. Because of the reactivities of silicon and most metals, this can dramatically limit the choices of materials.

In addition to having to remain intact to keep the silicon and the metal from interacting, the barrier must also keep either one of them from diffusing through and reacting with the other. Bulk diffusivities tend to scale with the melting point (*7*), so materials with a high melting point are preferred. In addition, it is important to consider high diffusivity paths, such as grain boundaries (*7*). If a single grain boundary crosses the thickness of the barrier, it may provide a ready diffusion pipe for the silicon or the metal, bypassing the bulk of the barrier. It is almost impossible to grow single crystal layers of materials in microelectronics processing, so eliminating grain boundaries requires using an amorphous material. Thus, an amorphous refractory material is a obvious candidate for a diffusion barrier.

A third set of demands on the diffusion barrier is related to the deposition method. There are two common methods for depositing films in microelectronics. The first is sputtering, where a target material is eroded by a plasma and transported to the silicon wafer, possibly reacting with the plasma gases along the way. The second method is chemical vapor deposition (CVD), where gas or vapor phase precursors flow over the wafer and react to grow the film. Sputtering has the advantage of being a low temperature, simple process. Since the diffusion barrier is deposited after the transistors are formed and, possibly, after some metal layers have already been deposited, the allowable thermal budget is very small. However, CVD processes are more effective at coating the sides or bottoms of vias, leading to better step coverage. This issue of step coverage is critical since via sizes are rapidly approaching 0.18 μm with an aspect ratio of 10:1 or greater.

Finally, several electrical parameters are required of the barrier. The first is that the barrier material conduct electricity. The requirement is not as stringent as for the metal wires, since the barriers are much thinner than the length of the wiring. So, although a resistivity of 2.4 μΩ cm for Al wires is considered high relative to 1.6 μΩ cm for Cu, it is generally accepted that future diffusion barriers can have resistivities up to 1000 μΩ cm and still be useful. The second electrical parameter is the contact resistance of the barrier to the silicon or the metal. Contact resistance is caused by the presence of an interface (and, possibly, contamination) between materials. A desirable contact resistance for future barriers is 1 nΩ/cm^2. For a square 0.15 μm contact, the contribution to the resistance per contact from a bulk resistivity of 1000 μΩ cm and a contact resistance of 1 nΩ/cm^2 are equal, and as the contact size decreases, the contact resistance rapidly becomes larger than the bulk resistance.

Currently, the most widely used diffusion barrier for microelectronics metallization is sputtered TiN. As device sizes continue to shrink, two major problems arise with sputtered TiN: poor step coverage and the columnar polycrystalline microstructure. The poor step coverage will not allow TiN to fully line the aggressive via geometries of future devices, while the columnar microstructure will limit the barrier performance by providing grain boundaries that act as fast diffusion paths across the barrier (1). Chemical vapor deposition of TiN has the potential to produce highly conformal films. Films grown from TiCl$_4$ and NH$_3$ exhibit excellent step coverage, but are deposited at relatively high temperatures (450 to 700°C) and are contaminated with chlorine, particularly at low temperatures, which can cause corrosion in the aluminum metallization (8). The deposition temperature can be reduced by using metalorganic precursors such as tetrakis(dimethylamido)titanium (TDMAT), with or without NH$_3$, but these films generally exhibit either poor step coverage or film resistivities that are too high for metallization applications (typically caused by C impurities) (9-11). More recent work has focused on improving the quality of films grown with TDMAT by post-deposition treatments with nitrogen plasmas or silane exposure (12, 13). Titanium nitride films grown from tetrakis(diethylamido)titanium (TDEAT) and NH$_3$ exhibit good step coverage and lower resistivities (11). Regardless of the deposition chemistry, the problem of grain boundary diffusion remains for CVD TiN.

To eliminate grain boundary diffusion, several groups have investigated amorphous or nanocrystalline materials (14-17). The most promising materials for microelectronics are films consisting of a refractory metal (e.g., Ti or W), nitrogen, and a third constituent, typically Si, such as titanium-silicon-nitride (Ti-Si-N) (16). These ternaries have high crystallization temperatures, so they retain their microstructure after thermal cycling, and tend to be thermodynamically stable in contact with either Si or the metal lines. Excellent barrier properties have been demonstrated for sputtered Ti-Si-N, W-Si-N, and other similar combinations (*16-18*). However, the poor step coverage of the sputtering process remains an obstacle to their application in ULSI devices.

In this report we describe the development of chemical vapor deposition (CVD) processes for three refractory ternary nitrides, Ti-Si-N, W-B-N, and W-Si-N. These materials do not react with either Si or Cu (*19*), and are either amorphous or contain small nanocrystals in an amorphous matrix, and do not crystallize at normal BEOL processing temperatures. The CVD processes described here are performed at low temperatures, generally near 350°C, and the step coverages for the processes are sufficient for future device geometries. The materials have resistivities below 1000 $\mu\Omega$ cm, suitable for use in metallization. These characteristics make them viable candidates for ULSI diffusion barriers.

Experimental

Titanium-silicon-nitride films were grown in an MRC Phoenix CVD system. This system consists of a single-wafer process chamber attached to a central robotics hub with a vertical-cassette elevator for batch wafer loading. The system was based on MRC's beta-test version of their $TiCl_4$ TiN CVD system, with modifications to deliver metalorganic precursors as well as an additional mass flow controller for silane. The metalorganic precursors were delivered using a heated (65°C) bubbler and N_2 as the carrier gas. The system at Sandia used nitrogen (N_2), ammonia (NH_3), silane (SiH_4), and a metalorganic precursor in nitrogen carrier gas. The only metalorganic precursor used in these experiments was TDEAT. Experiments performed using $TiCl_4$ are detailed elsewhere (*20*).

The MRC process chamber was designed to be similar to a rotating disc reactor (*8*). The 150 mm wafers sit on a susceptor that can be heated (20 - 800°C) and can rotate at 0 to 1500 rpm. All depositions were performed at a pressure of 20 Torr. The walls were warmed to 125-150°C to prevent condensation of TDEAT. A gas injection showerhead lies $\approx$3 cm above the susceptor. As delivered, all the reactants were injected into the showerhead plenum, allowed to mix, and then flowed down out of the showerhead to the wafer below. The showerhead, although passively cooled with a heat pipe, was warm enough that TDEAT would react with the other precursors while still in the showerhead. This resulted in the growth of highly C contaminated films. Therefore, the showerhead was modified to include a separate "piccolo tube" for injection of TDEAT. This tube hung slightly below the showerhead (which was moved up to accommodate the modifications) and extended along a radius from near the center of the showerhead out to the edge of the wafer.

Tungsten-based films were grown in a Vactronics CVD system, a single wafer, manual loading tool. The 150 mm wafer holder was heated to 300 - 450°C. The showerhead was custom designed and included three separately plumbed injection rings located 10 cm above the wafer. The gases used in film deposition were WF_6, NH_3, N_2, Ar, and a pyrophorics line used for B_2H_6, SiH_4, or Si_2H_6. The deposition pressure was typically 500 to 700 mTorr.

The weight gain during deposition was correlated with cross section SEM measurements of film thickness. These results were used to estimate the film thicknesses of other samples by using only the mass gain of the wafer. The sheet resistance of films deposited on oxide was measured using a 4-point probe, and converted to film resistivity using the estimated film thickness. Plan view or cross section high resolution TEM was performed on selected Ti-Si-N, W-B-N, and W-Si-N films to determine the microstructure of the material. Rocking curve X-ray data was taken on various films before and after thermal annealing to monitor microstructure changes. Film composition was measured with 3.5 MeV He^+ Rutherford backscattering spectrometry (RBS) and 28 MeV Si^{5+} elastic recoil detection (ERD) (21). Film stress was determined from wafer curvature measurements.

Ternary films were deposited over test structures to determine step coverage. These structures consisted either of trenches patterned through polysilicon down to silicon dioxide (for feature sizes down to 0.5 μm), or trenches were cut into crystal silicon and then thermally oxidized to narrow the trenches down to feature sizes of 0.1 μm. Films were also deposited over large area (250 × 250 μm) diodes for testing their effectiveness as diffusion barriers. Finally, several films were used in contact chain structures to monitor integration effects.

Titanium-Silicon-Nitride from TDEAT

Titanium nitride can be deposited by CVD at temperatures down below 250°C with metalorganic precursors such as TDMAT or TDEAT (22). We chose to focus on using TDEAT as the titanium precursor. TDMAT reacts more easily than TDEAT, which leads to two problems. The first is that TDMAT and NH_3 react quickly in the gas phase (23), which can lead to gas phase nucleation of particles. The reaction rate for TDEAT is more than two orders of magnitude lower than for TDMAT (24). Second, silane and disilane are much less reactive at low temperatures, so Si incorporation will be harder at lower temperatures where TDMAT is commonly used. This suggested that the deposition temperature might have to be relatively high by metalorganic CVD standards. However, it is known that TDMAT-based TiN can have high carbon levels, increasing the film resistivity, as the deposition temperature rises above 300°C (11). These constraints limited the choice to TDEAT as the titanium precursor.

Titanium-silicon-nitride films can be grown under a wide range of conditions using TDEAT, NH_3, and SiH_4 (25). The film compositions that have been deposited from 300 to 450°C are shown in Figure 2 on the Ti-Si-N ternary phase diagram (26). At these deposition temperatures, this chemistry can incorporate a wide range of Si in the films (0 to >25 at.%). The Ti-Si-N films lie along the TiN-Si_3N_4 tie line and are generally N rich. This is not surprising as MOCVD TiN is also found to be N rich (11, 23). Since TiN is known to have a broad composition range from 30 to over 54 at.% N (27, 28), the TiN-Si3N4 "tie line" is actually a two phase region that may encompass all of these films.

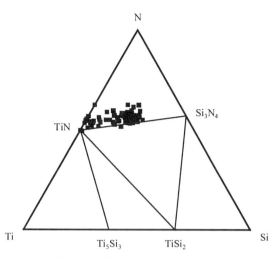

Figure 2. The compositions of films grown from TDEAT, SiH_4, and NH_3 are shown on a Ti-Si-N ternary phase diagram. The films lie near the TiN - Si_3N_4 tie line, and are generally nitrogen rich. The phase equilibria are tentative, since they are not known at these temperatures.

The question arises as to how Si is incorporated in the Ti-Si-N films. It is well known that SiH_4 and NH_3 do not react with each other at these low temperatures. This leaves three possibilities. The first is that the silane reacts with TDEAT, although TDEAT and SiH_4 do not react without NH_3 being present, i.e., we have not been able to deposit titanium silicides using these precursors. However, the reaction of silane may be catalyzed by an intermediate in the reaction of TDEAT and NH_3. TDEAT undergoes a transamination reaction in which the diethylamine groups surrounding the central Ti atom are replaced with NH_2 from the ammonia (*29, 30*). The more compact amine may reduce stearic hindrance *(31)*, and allow the SiH_4 access to the Ti atom to undergo the reaction. A second reaction path is that the decomposition of SiH_4 is catalyzed by a metallic surface. That is, a thin film of TiN grows first, and then SiH_4 could decompose on the surface and be incorporated in the growing film. The third possibility is that diethylamido complexes with silicon are formed in either the gas phase or on the surface. These complexes react to form Si_3N_4 at lower temperatures than required for the SiH_4-NH_3 reaction *(32)*, but it is not known if they will react below 450°C. Further experiments, such as mass spectrometry of the exhaust gases of the system, are needed to identify the Si reaction mechanism.

Although the mechanism for incorporation of Si in the Ti-Si-N films is not known, the amount of Si found in the films is a well behaved function of precursor flows. As shown in Figure 3, the relative concentration of Si found in a film is a monotonically increasing function of the ratio of SiH_4 to TDEAT in the precursor flow, divided by the total gas flow in the system. A similar plot is obtained for the simpler cases of either the ratio of SiH_4 to TDEAT or SiH_4 to total gas flow on the

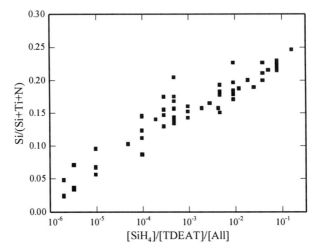

Figure 3. The Si content of Ti-Si-N is shown vs the ratio of SiH_4 to TDEAT divided by the total gas flow, which establishes the gas residence time in the reactor. Si incorporation increases logarithmically with the gas flow ratios. The spread is caused by changing parameters such as the temperature.

abscissa. The increase in Si content with increasing SiH_4 flow is expected. The decrease in Si content as the total gas flow is increased, and hence the SiH_4 residence time in the reactor decreases, is also normal. An increase in TDEAT flow could result in a decrease in the Si content if the major effect of higher TDEAT flows is to grow more TiN, lowering the relative amount of Si in the film. Two other points should be noted. First, very low SiH_4 flows are required to incorporate some Si into the films (typical SiH_4 flow rates where below 1 sccm). Second, the incorporation of Si into the films saturates relatively quickly with SiH_4 flow. That is, we could not access compositions near the Si_3N_4 side of the tie line.

Figure 4 shows the resistivity of Ti-Si-N films as a function of the Si content. The resistivity of TiN (no Si) grown from TDEAT and NH3 ranges from about 100 to 1000 $\mu\Omega$ cm. This spread is normal for this deposition chemistry, and reflects a range of TiN stoichiometries (33). The addition of silicon to the films increases the resistivity exponentially, eventually exceeding 1 Ω-cm at 25 at.% Si (not shown). In order to remain below the 1000 $\mu\Omega$-cm level, the films must have a composition with less than approximately 4 at.% Si.

The lowest resistivity at a given Si composition is found for films deposited at 350°C. Here, the film resistivity varies predictably from $\approx$ 400 $\mu\Omega$ cm for TiN to nearly 1 Ω cm for Ti-Si-N with 25 at.% Si. Films grown at 300, 400, or 450°C have higher resistivities at a given Si composition. These systematic variations in film resistivity with deposition temperature correlate well with the "excess" nitrogen incorporated in the films, that is, how far the films lie from the ideal TiN-Si_3N_4 tie line. Films deposited at 350°C have the lowest resistivities and lie closest to the tie line, having an average of only 1.5 at. % excess N. Films deposited at 300 or 400°C

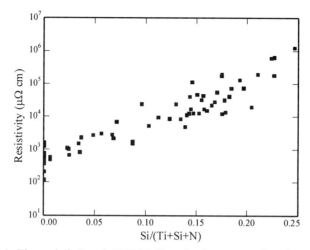

Figure 4. The resistivity of Ti-Si-N films is shown as a function of the Si content. The wide spread in resistivity at a given Si content is caused primarily by variations in the N content of the films, i.e. how far the films lie from the ideal TiN - Si_3N_4 tie line. Only films with low Si content have low resistivities.

have $\approx$ 8 at.% excess N, resulting in higher resistivities. The 450°C films have the greatest amount of excess N, $\approx$ 15 at.%, and the highest resistivities. This is similar to MOCVD TiN films deposited at atmospheric pressure, which have resistivities that are sensitive to stoichiometry and display an increase in resistivity at temperatures above and below 350°C (*33*).

High resolution cross-section TEM of Ti-Si-N show that, as-deposited, the film consists of small nanocrystals, $\approx$ 6 nm in diameter, embedded in an amorphous matrix (*25*). The rings in the electron diffraction pattern index to TiN, with no evidence of any other phase. Therefore, the nanocrystals observed are TiN embedded in an amorphous matrix. The volume fraction of amorphous material is much larger than can be accounted for by Si_3N_4 alone, so the amorphous material must also contain a significant fraction of Ti as well. The microstructure found for these CVD Ti-Si-N films is similar to that of sputtered Ti-Si-N films, which are known to be excellent diffusion barriers (*16*). X-ray diffraction indicates that annealing below 750°C does not significantly change the microstructure.

The Ti-Si-N films are smooth and featureless as deposited, and have good step coverage (*25*). For example, a $Ti_{0.46}Si_{0.03}N_{0.51}$ film (with impurities of 1 at.% C and 8 at.% H) deposited at 350°C has step coverages of 60% on 0.2 μm lines with an aspect ratio of 6:1, 75% on 0.35 μm lines at 3:1, and 35% to 40% even for 0.1 μm lines at 10:1. It should be noted that the conformality of this CVD Ti-Si-N film is better than that previously observed for optimized TDEAT-based CVD TiN films (85% for 0.8 μm lines with an aspect ratio of 1.2:1) (*11*).

Ti-Si-N films were integrated into a standard MDL test chip, a 16 kilobyte static random access memory surrounded by various test structures. A twelve wafer lot was split three ways at the via liner process. One split received the standard via liner, consisting of sputtered Ti, a rapid thermal anneal (RTA) step to resilicide the contacts and getter impurities into the Ti, and sputtered TiN. The second split was processed through sputtered Ti, RTA, and then CVD Ti-Si-N. The final split consisted of using just CVD Ti-Si-N alone with no sputtered Ti. It was found that CVD tungsten nucleated differently on the CVD Ti-Si-N than on sputtered TiN.

After completing the rest of the needed processing (tungsten deposition, CMP, and metal deposition and patterning), all wafers went through electrical test. The resistivity of chains of 2986 contacts from metal 1, through a tungsten plug (with Ti/TiN or CVD Ti-Si-N liner) to n+ Si, and then back up through a plug to metal 1 were measured to determine the average resistance per contact. The four wafers with the standard Ti/TiN process yield an average resistance per contact of $13.6 \pm 3.6 \ \Omega$ (251 total good die). Only one CVD Ti-Si-N wafer survived processing because of tungsten nucleation problems, and it had an average resistance per contact of $13.8 \pm 4.0 \ \Omega$ (63 good die). The mixed process, sputtered Ti and then CVD Ti-Si-N, showed much higher resistances of about 56 Ω, which was attributed to the exposed Ti absorbing impurities, primarily oxygen, while waiting for the CVD Ti-Si-N film. Comparing the standard Ti/TiN process vs the CVD Ti-Si-N liner shows no statistically significant difference in contact resistance.

Several different Ti-Si-N films were evaluated as diffusion barriers against copper or aluminum. Arrays of large area diodes ($250 \times 250 \ \mu m$) isolated by oxide were fabricated on 150 mm wafers. After depositing a blanket barrier film, the Ti-Si-N was patterned so that it covered the diode area and some 10 μm of oxide on all sides. Thick photoresist was used to pattern sputtered copper or aluminum using a lift-off process. The diodes could then be annealed in vacuum at different temperatures, and the diode characteristics, particularly the reverse leakage current, used as a measure of barrier effectiveness.

With Al metallization, all of the Ti-Si-N barriers survived anneals up to 30 min at 350°C, but all of them fail after annealing at 400°C. Experiments on sputtered Ti-Si-N barriers against Al typically perform better than these CVD films (*34*). However, all of the sputtered films that were tested had higher Si contents than these CVD Ti-Si-N films. For Cu metallization, no increase in reverse current, indicative of barrier failure, is observed for anneals up to 500°C. After annealing at 550°C, most barriers are still good, and it is only at 600°C that all barriers fail. This performance should be suitable for use in future ULSI generations.

Tungsten-Boron-Nitride

Tungsten-boron-nitride films were deposited in the Vactronic CVD system (*35*). The precursors used for the film were WF_6, SiH_4, NH_3, and 30 % B_2H_6 in N_2. Argon was used as a carrier gas. Tungsten is incorporated in the film through reduction of WF_6, which thermodynamically prefers to react with the silane. Because B_2H_6 is unstable at room temperature, boron is most likely incorporated in the film through simple decomposition of B_2H_6. Nitrogen is incorporated through the reaction of ammonia with tungsten on the hot wafer surface (*36*). The only impurities detected by RBS or ERD in these films was Si (less than 5 at.% maximum) and H (typically 5 at.%).

The compositions grown with this chemistry are shown in Figure 5 on a W-B-N ternary phase diagram. Since the ternary phase diagram has only been assessed at much higher temperatures (26), the tie lines shown are tentative. The main discrepancy with published diagrams is the possible tie lines between BN and the different tungsten nitrides, many of which are thermodynamically unstable (37), but are easily grown by CVD. The compositions lie roughly in a band from WB_4 to W_2N. We were unable to increase the tungsten fraction in the film by changing the relative gas flows. Films grown with high WF_6 partial pressures have little or no detectable hydrogen in them, compared with 5 at.% H for most of the films shown on the diagram. This indicates that for sufficiently high WF_6 flows the growing surface switches from being H-terminated to being F-terminated, which has been observed in the WF_6-SiH_4 system (38). This change in the surface state may limit W incorporation in the film.

As-deposited W-B-N films are amorphous as measured by x-ray diffraction. The resistivities of these films range from 200 $\mu\Omega$ cm for films rich in tungsten and boron to 20,000 $\mu\Omega$ cm for films close to the W-N side of the ternary system. The resistivity as a function of N content is shown in Figure 6. Films with less than about 20 at.% N have resistivities below 1000 $\mu\Omega$ cm.

The step coverage of a $W_{0.24}B_{0.66}N_{0.10}$ film grown over 1.5 μm wide trenches with an aspect ration of 5.5 is 40%, which is good but not exceptional. However, the large surface area caused by having a high density of these structures across the entire wafer may have resulted in reactant depletion which would limit the coverage. It is possible that the step coverage on samples more typical of microelectronics metallization might be better.

All ternary W-B-N films investigated with Read camera x-ray diffraction

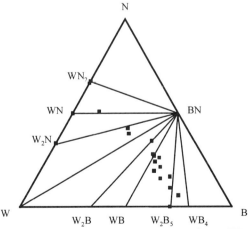

Figure 5. The compositions of films grown at 350°C from WF_6, B_2H_6, and NH_3 (with SiH_4 to reduce the WF_6) are shown on a W-B-N ternary phase diagram.

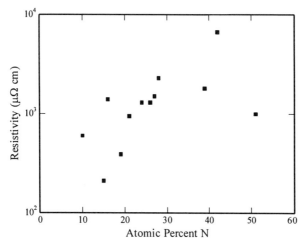

Figure 6. The resistivity of W-B-N films is shown as a function of the N content. Several different compositions have resistivities below 1000 $\mu\Omega$ cm, required for use as diffusion barriers.

were amorphous as deposited. High resolution TEM (not shown) on a film with the composition of $W_{0.24}B_{0.52}N_{0.24}$ revealed some small diffracting regions some 1.5 to 2 nm in size. However, the volume fraction of these crystallites was estimated to be below 5 vol.%. Films were then subjected to one hour anneals until some crystallization was detected in the x-ray diffraction patterns. Many compositions did not begin to crystallize until 700°C or higher.

These films were also tested as diffusion barriers using 100 nm films deposited on 250 × 250 μm diodes. These diodes had Cu sputtered on them, and the diode characteristics were monitored as a function of anneal temperature in vacuum. Failure of the barrier was defined as an increase in reverse current of a factor of ten. All of these films were good diffusion barriers up to at least 500°C for 30 min, with many composition exceeding 600°C.

Tungsten-Silicon-Nitride

Tungsten-silicon-nitride films were also deposited in the Vactronics CVD system (35). The precursors used were WF_6, SiH_4 or Si_2H_6, and NH_3. Argon was used as a diluent. The deposition pressure was typically 500 to 700 mTorr, and the wafer temperature was 350°C. Little if any silicon was incorporated in the film when SiH_4 was used. However, using Si_2H_6 did result in Si incorporation. The deposition rates are adequate for diffusion barrier applications at 5 to 500 nm/min, depending on flow conditions. The films were smooth and featureless both visually and by SEM. The only impurity detected in the films was H, present at levels below 5 at.%.

A range of compositions can be grown from WF_6, Si_2H_6, and NH_3 as indicated in Figure 7. In general, the compositions form an arc from WN_2 to WSi_2, passing over a number of possible tie lines. As in the cases of Ti-Si-N and W-B-N, the ternary phase diagram has only been assessed at much higher temperatures (*26*), so the tie lines shown are tentative, particularly for the W-N compounds. The film resistivity as a function of Si content is shown in Figure 8. Almost all films have resistivities below 1000 $\mu\Omega$ cm. Films with stoichiometries corresponding to WN_2 (i.e. no silicon) have resistivities in excess of 20,000 $\mu\Omega$ cm.

The step coverage of these films is exceptional. Even over reentrant 0.25 μm, 4:1 aspect ratio features the step coverage is 100%. Films with a composition of $W_{0.47}Si_{0.09}N_{0.44}$ have been tested as diffusion barriers with copper on large area diodes. The reverse leakage current of the diodes increased only after a 30 min anneal at 700°C. This film composition shows signs of crystallization by x-ray diffraction at 700°C as well. High resolution TEM on as-deposited $W_{0.49}Si_{0.10}N_{0.41}$ films shows no evidence for crystallinity in bright field, dark field, or electron diffraction imaging modes.

Conclusions

Processes for chemical vapor deposition of three different amorphous refractory ternary materials have been developed. Titanium-silicon-nitride was deposited using TDEAT, NH_3, and SiH_4. The Ti-Si-N films from TDEAT consisted of nanocrystals of TiN in an amorphous matrix. Growth of tungsten-boron-nitride was accomplished with a mixture of WF_6, SiH_4, B_2H_6, and NH_3, while tungsten-silicon-nitride was deposited from WF_6, Si_2H_6, and NH_3. The tungsten-based ternaries were completely amorphous as-deposited. Each film offers step coverages and film resistivities well within the range needed for future ULSI generations. In addition, they are excellent diffusion barriers to copper, with barrier failure temperatures as high as 700°C. These characteristics make them promising candidates for advanced diffusion barriers in microelectronics applications.

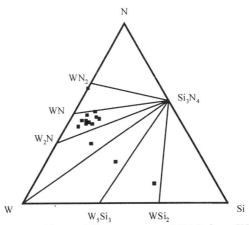

Figure 7. The compositions of films grown at 350°C from WF_6, Si_2H_6, and NH_3 are shown on a W-Si-N ternary phase diagram. The tie lines are tentative. Most films lie in a composition range from WSi_2 to WN.

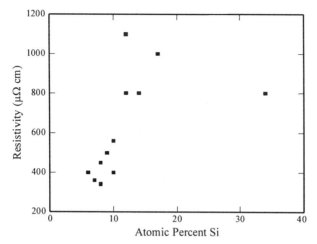

Figure 8. The resistivity of W-Si-N films is shown as a function of the Si content. Most films have resistivities below 1000 μΩ cm, required for use as diffusion barriers.

These experiments were performed at the Microelectronics Development Laboratory (MDL) at Sandia. Sandia is a multiprogram laboratory operated by Sandia Corporation, a Lockheed Martin Company, for the United States Department of Energy under Contract DE-AC04-94AL85000.

References

1. Nicolet, M.-A., Thin Solid Films **1978**, 52, 415.
2. Ho, P. S., Thin Solid Films **1982**, 96, 301.
3. Nowicki, R. S.; Nicolet, M.-A. Thin Solid Films **1982**, 96, 317.
4. Ting, C. Y.; Wittmer, M. Thin Solid Films **1982**, 96, 327.
5. Wittmer, M. J. Vac. Sci. Technol A **1984**, 2, 273.
6. Murarka, S. P. Defect and Diffusion Forum **1988**, 59, 99.
7. Porter, D. A.; Easterling, K. E. *Phase Transformations in Metals and Alloys 2nd edition*; Chapman & Hall: New York, NY, 1992.
8. Studiner, D. W.; Hillman, J. T.; Arora, R; Foster, R. F. In *Advanced Metallization for ULSI Applications 1992*, Cale, T. S.; Pintchovski, F. S., Eds.; Materials Research Society: Pittsburgh, PA, 1993; pp. 211-217.
9. Cale, T. S.; Raupp, G. B.; Hillman, J. T.; Rice, Jr., M. J. In *Advanced Metallization for ULSI Applications 1992*, Cale, T. S.; Pintchovski, F. S., Eds.; Materials Research Society: Pittsburgh, PA, 1993; pp. 195-202.
10. Eizenberg, M.; Littau, K.; Ghanayem, S.; Mak, A.; Maeda, Y.; Chang, M.; Sinha, A. K. Appl. Phys. Lett. **1994**, 65, 2416.
11. Raaimakers, I. J. Thin Solid Films **1994**, 247, 85.
12. Danek M.; Liao M.; Tseng J.; Littau K.; Saigal D.; Zhang H.; Mosely R.; Eizenberg M. Appl. Phys. Lett. **1996**, 68, 1015.
13. Lu J. P.; Hsu W. Y.; Hong Q. Z.; Dixit G. A.; Luttmer J. D.; Havemann R. H.; Magel L. K. J. Electrochem. Soc. **1996**, 146, L279.

14. Doyle B. S.; Peercy, P. S.; Wiley, J. D.; Perepezko, J. H.; Nordman, J. E. J. Appl. Phys. **1982**, 53, 6186.
15. Saris, F. S.; Hung, L. S.; Nastasi, M.; Mayer, J. W.; Whitehead, B. Appl. Phys. Lett. **1985**, 46, 646.
16. Reid, J. S.; Sun, X.; Kolowa, E.; Nicolet, M.-A. IEEE Elec. Dev. Lett. **1994**, 15, 298.
17. Reid, J. S.; Kolawa, E.; Garland, C. M.; Nicolet, M.-A.; Cardone, F.; Gupta, D.; Ruiz, R. P. J. Appl. Phys. **1996**, 79, 1109.
18. Asai, K.; Sugahara, H.; Matsuoka, Y.; Tokumitsu, M. J. Vac. Sci. Tech. **1988**, B6, 1526.
19. J. S. Reid, E. Kolowa, and M.-A. Nicolet, J. Mater. Res. **1992**, 7, 2424.
20. Smith, P. M.; Custer, J. S.; Jones, R. V.; Maverick, A. W.; Roberts, D. A.; Norman, J. A. T.; Hochberg, A. K.; Bai, G.; Reid, J. S.; Nicolet, M.-A. In *Advanced Metallization and Interconnect Systems for ULSI Applications in 1995*, Ellwanger, R. C.; Wang, S.-Q., Eds.; Materials Research Society: Pittsburgh, PA, 1996; p. 249.
21. *Handbook of Modern Ion Beam Materials Analysis*; Tesmer, J. R.; Nastasi, M., Eds.; Materials Research Society: Pittsburgh, PA, 1995.
22. Fix, R.; Gordon, R. G.; Hoffman, D. M. Chem. Mater. **1991**, 3, 1138.
23. Weiller, B. H.; Partido, B. V. Chem. Mater. **1994**, 6, 260.
24. Weiller, B. H. Chem. Mater. **1995**, 7, 1609.
25. Smith, P. M; Custer, J. S. Appl. Phys. Lett. **1997**, 70, 3116.
26. Rogl P.; Schuster, J. C. *Phase Diagrams of Ternary Boron Nitride and Silicon Nitride Systems*; ASM International: Materials Park, OH, 1992.
27. Wriedt H. A.; Murray, J. L. Bull. Alloy Phase Diag. **1987**, 8, 378.
28. Ohtani, H.; Hillert, M. Calphad **1990**, 14, 289.
29. Prybyla, J. A.; Chiang, C.-M.; Dubois, L. H. J. Electrochem. Soc. **1993**, 140, 2695.
30. Weiller, B. H. Mat. Res. Soc. Symp. Proc. **1994**, 335, 159.
31. Bradley, D. C.; Thomas, I. M. J. Chem. Soc., **1960**, 3857.
32. Gordon, R. G.; Hoffman, D. M.; Riaz, U. Chem. Mater. **1990**, 2, 480.
33. Musher, J. N.; Gordon, R. G. J. Electrochem. Soc. **1996**, 143, 736.
34. Sun, X.; Reid, J. S.; Kolawa, E.; Nicolet, M.-A.; Ruiz, R. P. J. Appl. Phys. **1997**, 81, 664.
35. Fleming, J. G.; Smith, P. M.; Custer, J. S.; Roherty-Osmun, E.; Cohn, M.; Jones, R. V.; Roberts, D. A.; Norman, J. A. T.; Hochberg, A. K.; Reid, J. S.; Kim, Y.-D.; Kacsich, T.; Nicolet, M.-A. In *Advanced Metallization and Interconnect Systems for ULSI Applications in 1996*, Havemann, R.; Schmitz, J.; Komiyama, H.; Tsubouchi, K., Eds.; Materials Research Society: Pittsburgh, PA, 1997; p. 245.
36. *Gmelin Handbook of Inorganic Chemistry, 8th edition*; Springer-Verlag: New York, NY, 1987.
37. Wriedt, H. A. Bull. Alloy Phase Diag. **1989**, 10, 358.
38. Yu, M. L.; Eldridge, B. N.; Joshi, R. V. In Tungsten and Other Refractory Metals for VLSI Applications I; Blewer, R. S.; McConica, C. M., Eds.; Materials Research Society: Pittsburgh, PA, 1989; p. 221.

Chapter 8

New Routes Toward Chemical and Photochemical Vapor Deposition of Copper Metal

Andrew W. Maverick, Alicia M. James, Hui Fan, Ralph A. Isovitsch, Michael P. Stewart, Ezana Azene, and Zuzanna T. Cygan

Department of Chemistry, Louisiana State University, Baton Rouge, LA 70803-1804

Several types of volatile copper(I) and copper(II) complexes have been studied as precursors for chemical vapor deposition (CVD) of Cu metal films. Adducts of $Cu(hfac)_2$ (hfacH = hexafluoroacetylacetone) with alcohols ROH (R = Me, Et, n-Pr, i-Pr, n-Bu, s-Bu, i-Bu) are self-reducing: CVD occurs by reduction of copper(II) to Cu metal, while the alcohols are oxidized to the corresponding carbonyl compounds. The best results have been obtained for $Cu(hfac)_2 \cdot i$-PrOH (1.3 μm h^{-1} under N_2 at 200 °C, vs. 0.46 μm h^{-1} under H_2). The adduct of $Cu^{II}(hfac)_2$ with hydrazine is also self-reducing, leading to Cu metal formation under N_2 at substrate temperatures of 140 °C. Adducts with other nitrogen bases (e.g. piperidine) can be used for Cu CVD with H_2 carrier gas. The copper(I) amide cluster $[CuN(SiMe_3)_2]_4$ can be used for Cu CVD under H_2. It is also phosphorescent, and it shows a slight lowering of the threshold substrate temperature for Cu deposition under UV irradiation; this suggests that the deposition can be photochemically induced.

Introduction

One of the dominant trends of the semiconductor industry is toward integrated circuits with smaller transistors more densely packed together. The higher packing densities generally permit greater operating speeds, and the increasing number of devices per chip leads to greater functionality per unit cost. As device sizes decrease into the sub-micron range, the performance of the aluminum and tungsten interconnects now used to make connections between devices is likely to deteriorate, and this may cause serious problems in speed and reliability. One of the most promising alternative materials for interconnects is copper: it is superior to both Al and W in resistivity, and superior to Al in resistance to electromigration and other types of failure (*1*). Chemical vapor deposition (CVD) is an attractive method for production of Cu films (like many other materials used in IC manufacture), because it can lead to films of high purity that cover surfaces of a variety of different topographies on the chip.

Both copper(II) and copper(I) precursors have been used for CVD of copper metal (2). Complexes of hexafluoroacetylacetonate (hfac⁻) and related ligands have dominated both families, with the most commonly used precursors being $Cu^{II}(hfac)_2$ and $Cu^I(hfac)L$ (L = neutral ligand, e.g. alkene, alkyne, CO, phosphine). With the copper(II) precursors, deposition of metal has generally been carried out by reduction using H_2 as carrier gas:

$$Cu^{II}(hfac)_2 + H_2 \longrightarrow Cu(s) + 2hfacH(g) \qquad (1)$$

Copper(I) precursors can also be reduced by H_2 (3), but they are especially well suited to disproportionation (eq 2), because the oxidized product, $Cu(hfac)_2$, is also volatile:

$$2Cu^I(hfac)L \longrightarrow Cu(s) + Cu^{II}(hfac)_2(g) + 2L(g) \qquad (2)$$

Among the most promising copper(I) precursors are Cu(hfac)(alkene) (alkene = 1,5-cyclooctadiene (4), 1,5-dimethyl-1,5-cyclooctadiene (5), and trimethylvinylsilane (6)). These precursors have shown some of the highest Cu deposition rates reported with any copper system (7).

We report here our recent progress in two aspects of copper CVD. First, we have used self-reducing copper(II) precursors, containing reducing ligands bound to $Cu(hfac)_2$. Many of these are liquids at typical precursor evaporation temperatures (80-100 °C), and some show significantly higher Cu deposition rates than traditional copper(II) precursors. Second, we have explored copper(I) precursors with low-energy excited states, with the goal of using photochemical methods for deposition of Cu films.

Results and Discussion

Chemistry of Cu(hfac)₂ and its use in CVD. $Cu(hfac)_2$ is a versatile Lewis acid, forming adducts with a variety of bases. Hydrates, $Cu(hfac)_2 \cdot xH_2O$, were among the first such adducts isolated, although the properties of the monohydrate and dihydrate (as compared with the anhydrous compound) were not well established until the late 1960s (8). Other adducts reported in the early literature include those with pyridine (9) and triethylamine (10). One indication of the generality of $Cu(hfac)_2$ as a Lewis acid is its inclusion as one of the very few transition-metal complexes in tables (of so-called E and C parameters) for the quantitative treatment of Lewis acid-base reaction thermodynamics (11).

Although these other adducts of $Cu(hfac)_2$ had been prepared, most of the experiments dealing with its reduction to Cu metal, including the early Cu CVD studies, used the anhydrous compound or the hydrate. In one of the first studies of the chemistry of $Cu(hfac)_2$ with other Lewis bases, Gafney and Lintvedt showed that UV irradiation of $Cu(hfac)_2$ in alcohol solutions (which contain $Cu(hfac)_2 \cdot xROH$) led to the deposition of Cu metal (12). In the mid-1980s, a group at IBM explored laser-induced CVD (13) of copper films using $Cu(hfac)_2 \cdot EtOH$ (14) as precursor. They found that using the alcohol adduct led to significantly smaller impurity concentrations (especially C and F) in the deposited films compared to the hydrate precursor. In analogy with the Gafney and Lintvedt experiments, they proposed that the alcohol was assisting by functioning as a reducing agent.

In the early 1990s, the addition of water and alcohols to Cu(hfac)$_2$ was studied as a method for improving deposition rates for conventional Cu CVD. Awaya and Arita showed that the addition of small amounts of H$_2$O accelerated Cu film deposition, whereas larger amounts led to deposition of copper oxide (*15*). Previous studies of conventional CVD from Cu(hfac)$_2$ in the presence of alcohols have been of several types. Cho (*16*) demonstrated enhanced Cu CVD when *i*-PrOH vapor was added to the gas mixture (H$_2$ + Cu(hfac)$_2$(g)) entering the reactor. Kaloyeros et al. (*17*) reported improved deposition rates for Cu(hfac)$_2$ via plasma-assisted CVD under H$_2$ when the precursor was dissolved in EtOH and *i*-PrOH and the solution vaporized into the carrier gas. Finally, Pilkington et al. showed that Cu(hfac)$_2$–ROH mixtures (ROH = EtOH or a propanol isomer) could be used to produce Cu films under N$_2$ carrier gas at higher temperatures (300-350 °C) (*18*).

Griffin and co-workers have proposed the following mechanism for Cu deposition via H$_2$ reduction of Cu(hfac)$_2$ (*19*). The rate-limiting step is believed to be reaction 5, the recombination of adsorbed H atoms and hfac groups to permit desorption of hfacH from the surface.

$$Cu(hfac)_2 \; \rightleftharpoons \; Cu(hfac)(ads) + hfac(ads) \tag{3}$$

$$H_2 \; \rightleftharpoons \; 2H(ads) \tag{4}$$

$$hfac(ads) + H(ads) \; \rightleftharpoons \; hfacH(g) \tag{5}$$

$$Cu(hfac)(ads) \; \longrightarrow \; hfac(ads) + Cu(s) \tag{6}$$

One possible rationale for the higher deposition rates observed in the presence of water and alcohols is that they serve as H sources by increasing the concentration of adsorbed H atoms on the metal surface; see reaction 7 below. When H$_2$ is present, the H atoms it supplies can later recombine with RO(ads) (i.e. reaction 7 is reversible), making the overall role of ROH essentially catalytic.

$$ROH(g) \; \rightleftharpoons \; RO(ads) + H(ads) \tag{7}$$

We were interested in the alcohol adducts for two reasons. First, in addition to the proton-transfer capabilities outlined above, alcohols are well known as reducing agents. Thus, it was possible that the alcohol molecules could reduce Cu(II) to Cu metal, thus making reducing carrier gases (e.g. H$_2$) unnecessary for Cu CVD. That is, primary and secondary alcohols ROH have a second possibility available other than recombination of RO(ads) + H(ads): loss of an additional H atom from the adsorbed alkoxy group to make an aldehyde or ketone. This process has been studied on a variety of surfaces (*20*).

The early work with Cu(hfac)$_2$·EtOH, both in solution and under CVD conditions (*12,13*), demonstrated this chemical function, but only under irradiation with UV light. In most of the other studies of Cu CVD with addition of alcohols, H$_2$ was also present, so the reducing properties of the alcohols in conventional CVD were not tested directly. The preliminary report of Pilkington et al. (*18*) appears to be the only previous example of conventional Cu CVD in which the added alcohol is likely to have served as reducing agent.

Our second reason for interest in the alcohol complexes was the low melting point (53-55 °C) of the adduct with 2-propanol, Cu(hfac)$_2$·*i*-PrOH (*14*). This means that at typical evaporator temperatures (80-100 °C), Cu(hfac)$_2$·*i*-PrOH is a liquid, which would make it easier to introduce it (via a bubbler or direct-liquid-injection system) into a CVD reactor in a controlled fashion.

Energetics. The ease of oxidation of 2-propanol to acetone is illustrated by the overall thermodynamic data in Table I (*21*) and by the bond dissociation energies in Table II (*22*). The enthalpy for oxidation of 2-propanol to acetone (Table I) is significantly more favorable than that for oxidation of methanol or other primary alcohols to aldehydes.

A related question that can be addressed by the bond dissociation energy data in Table II is whether individual steps in the alcohol oxidation reactions are likely to be faster for specific alcohols. The data show that cleavage of the C–H bond at the alpha carbon atom (i.e. closest to the OH group) is also substantially easier for 2-propanol than for the other alcohols, whereas dissociation of C–O and O–H bonds requires approximately the same amount of energy for all of the alcohols listed. Thus, if C–H bond cleavage is important in the rate-determining step in the deposition reaction, it can be expected to proceed faster when *i*-PrOH is the reductant.

If dissociative adsorption of alcohol on the substrate surface (reaction 4) is important in the CVD process, then the acidity values listed in Table III may be useful. Aqueous pK_a values are lowest for methanol and water, and highest for 2-propanol (*23*). (Autoprotolysis constants for the pure liquids show a similar trend (*24*).) Gas-phase acidities are in the opposite order, with 2-propanol *most* acidic (*25*). If the surface reaction is similar to the gas-phase process, then 2-propanol would be expected to dissociate most extensively on Cu, leading to the greatest acceleration of hfacH(g) desorption.

Table I. Thermodynamics of Alcohol Oxidation[a]

Reaction	ΔH_f (alc, g)	ΔH_f (ald/ ket, g)	ΔH_{rxn}
CH$_3$OH(g) → CH$_2$O(g) + H$_2$(g)	−201.0	−115.9	85.1
CH$_3$CH$_2$OH(g) → CH$_3$CHO(g) + H$_2$(g)	−235.3	−170.7	64.6
CH$_3$CH$_2$CH$_2$OH(g) → CH$_3$CH$_2$CHO(g) + H$_2$(g)	−255.6	−188.7	66.9
(CH$_3$)$_2$CHOH(g) → (CH$_3$)$_2$CO(g) + H$_2$(g)	−272.8	−218.5	54.3

[a]All values in kJ mol^{-1}. Data from (*21*).

Table II. Bond Dissociation Energies D_e for Alcohols[a]

Alcohol	D_e (H–C)	D_e (C–O)	D_e (O–H)
H–CH$_2$–O–H (methanol)	410	387	436
H–CH(CH$_3$)–O–H (ethanol)	389[b]	393	438[b]
H–CH(CH$_2$CH$_3$)–O–H (1-propanol)	ca. 387[c]	395	432[b]
H–C(CH$_3$)$_2$–O–H (2-propanol)	381[b]	402	438[b]

[a]All values in kJ mol^{-1}. Data from (*22*), except as noted. [b]Data from (*21*).
[c]Estimated from data for alcohols (shown here) and for ethane and propane.

Table III. Solution and Gas-Phase Acidity of Alcohols[a]

Alcohol	pK_a[a]	Gas-phase acidity[b]
CH_3OH (methanol)	15.5	1565 kJ mol^{-1}
CH_3CH_2OH (ethanol)	15.9	1555
$CH_3CH_2CH_2OH$ (1-propanol)	16.1	1544
$(CH_3)_2CH(OH)$ (2-propanol)	17.1	1545

[a]In H_2O, from (23). [b]Defined as ΔG for ROH(g) → RO$^-$(g) + H$^+$(g), from (25).

CVD with alcohol adducts. We have used several of the alcohol adducts as precursors in Cu CVD experiments, as summarized in Table IV. Deposition of Cu films proceeds smoothly in all cases when H_2 is the carrier gas; results are similar to those obtained when $Cu(hfac)_2 \cdot H_2O$ is used as the precursor. Under N_2 carrier gas, however, any Cu deposition requires a different reducing agent. In these experiments, we trapped the gaseous products and analyzed them by FTIR and GC/MS methods; for example, we detected formaldehyde, acetaldehyde, and acetone from the MeOH, EtOH, and i-PrOH precursors, respectively. Thus, we attribute the Cu film deposition under N_2 carrier gas to reduction of the Cu(II) precursor by the alcohol.

We have studied $Cu(hfac)_2 \cdot i$-PrOH most extensively (26). CVD experiments with this precursor under pure N_2 do not give copper films of good quality, whereas deposition under N_2 with excess i-PrOH vapor (introduced via a bubbler upstream of the

Table IV. CVD of Cu Films Using Cu(hfac)$_2$–Alcohol Adducts as Precursors[a]

Precursor	m. p. (°C)	Carrier gas[b]	Thickness/nm	Resistivity /$\mu\Omega$ cm
$Cu(hfac)_2 \cdot MeOH$	134-138[c]	N_2 + MeOH(g)	thin	
$Cu(hfac)_2 \cdot EtOH$	103-104[c]	N_2 + EtOH(g)	thin	
$Cu(hfac)_2 \cdot n$-PrOH	67-70	N_2 + n-PrOH(g)	200-500	2-3
$Cu(hfac)_2 \cdot i$-PrOH	53-55	N_2 + i-PrOH(g)	**1300**	2.9
"		H_2 + i-PrOH(g)	460	2.0
"		N_2	none	
"		H_2	280	2.8
$Cu(hfac)_2 \cdot n$-BuOH	66-75	N_2 + n-BuOH(g)	200-500	2-3
$Cu(hfac)_2 \cdot s$-BuOH	39-42	N_2 + s-BuOH(g)	300-600	2-2.5
$Cu(hfac)_2 \cdot t$-BuOH	< 40	N_2 + t-BuOH(g)	none	
$Cu(hfac)_2 \cdot i$-BuOH	50-55			
$Cu(hfac)_2 \cdot 2N_2H_4$	120 dec	N_2	> 100 (140 °C)	
$[CuN(SiMe_3)_2]_4$	> 200	H_2	ca. 20	high

[a]Standard conditions: carrier gas ca. 0.8 L min^{-1}, 1 atm; 1 h; evaporation temperature 80 °C; substrate (glass or Si) temperature 200 °C. [b]In most experiments (as indicated), alcohol vapor was added to carrier gas via a bubbler before passage through the evaporator. [c]Ref (14).

evaporation chamber) is much more successful. There are two possible reasons for this result. First, the adduct is known to be sensitive to displacement of coordinated alcohol by traces of water in the carrier gas or in the CVD apparatus. For example, the adducts can be sublimed without decomposition (as was originally reported for the MeOH, EtOH, and *i*-PrOH adducts (*14*)), and quadrupole mass spectra of Cu(hfac)$_2$–alcohol mixtures show fragments that still contain alcohol molecules coordinated to Cu (*27*). However, extended treatment with dry carrier gas (e.g. under typical precursor evaporator conditions) converts the alcohol adducts to anhydrous Cu(hfac)$_2$. Second, rapid reduction of the precursor by alcohol molecules on the CVD substrate may require that an excess of alcohol be present.

In addition to the adducts with MeOH, EtOH, and *i*-PrOH, we were interested in those with other alcohols. All of the alcohol adducts except those with MeOH and EtOH (see Table IV) have melting points sufficiently low for use as liquid precursors under typical CVD evaporator conditions. Also, several of the alcohols have higher boiling points (and correspondingly lower vapor pressures at room temperature); this might make their adducts more stable with respect to dissociation/ligand exchange. 1-Propanol is less volatile than 2-propanol; however, its adduct (shown in Table IV) also requires excess alcohol vapor for efficient reduction to Cu under N$_2$. More recently, we have obtained similar results with Cu(hfac)$_2$ adducts of 1- and 2-butanol and 2-methyl-1-propanol (isobutyl alcohol). Thus, the best CVD performance with all of these precursors is obtained in the presence of excess alcohol.

Cu(hfac)$_2$·*t*-BuOH cannot be used to deposit Cu under N$_2$, even when excess *t*-BuOH is present. We attribute this to the lack of α-H atoms in *t*-BuOH, which prevents it from functioning easily as a reducing agent.

Our successful Cu deposition using alcohols as reductants suggests the following reaction steps, in addition to reactions 3-6. With *i*-PrOH as the reductant, adsorption (eq 8) is likely to be followed by transfer of a second H atom from the adsorbed isopropoxy group to adsorbed hfac, with desorption of hfacH and acetone (eq 9).

$$(CH_3)_2CHOH \rightleftharpoons (CH_3)_2CHO(ads) + H(ads) \qquad (8)$$

$$hfac(ads) + (CH_3)_2CHO(ads) \rightleftharpoons hfacH(g) + (CH_3)_2C=O(g) \qquad (9)$$

Experiments designed to test this mechanism are now under way.

Amine and related N-adducts of Cu(hfac)$_2$. Adducts with amines are normally more strongly bound than those with alcohols, as would be expected due to their greater basicity. This was documented for Cu(hfac)$_2$ in work by Drago and co-workers (*11*): they measured binding constants of at least 1000 M^{-1} (i.e. Δ*G* ≤ ca. −17 kJ mol^{-1}) for Cu(hfac)$_2$ with pyridine and NEt$_3$ in solution at room temperature. More recently, Caulton and co-workers have reported the properties of Cu(hfac)$_2$·NH$_3$ (*28*), and its use for CVD of Cu$_3$N. They have also prepared several six-coordinate adducts of Cu(hfac)$_2$ with chelating amino alcohols (*29*). We wished to determine whether adducts could be prepared with monodentate nitrogen bases that were also good reducing agents. These may be usable as self-reducing Cu CVD precursors; also, their greater stability may make it unnecessary to use excess amine vapor in the CVD process.

Several oxidation (dehydrogenation) reactions are possible for nitrogen-containing compounds. In contrast to alcohol/ketone and alcohol/aldehyde systems, which have been extensively studied, only limited thermodynamic data are available for analogous reactions of amines. This is partly because the oxidized compounds, e.g. imines, are more difficult to isolate and purify than the corresponding carbonyl compounds. Enthalpies of dehydrogenation for methylamine, ethylamine, dimethyl-amine, and pyrrolidine (in the latter case making 1-pyrroline; see reactions 10 and 11) can be estimated to be 92, 71, 64, and 67 kJ mol^{-1} (30) (31). Dehydrogenation of several 5- and 6-carbon secondary amines, to produce imines (as in reaction 10), has been found to be endothermic by 80–96 kJ mol^{-1} (32). Thus, there do not seem to be any groups of simple amines that are likely to be as favorable as reductants for Cu(hfac)$_2$ (from a thermodynamic point of view) as i-PrOH.

$$RCH_2-NHR' \longrightarrow RCH=NR' + H_2 \qquad (10)$$

$$\qquad (11)$$

In contrast to normal amines, hydrazine and many of its derivatives are readily oxidized. For example, decomposition of gaseous hydrazine into N$_2$ and H$_2$ (reaction 12) is *exothermic* by 95 kJ mol^{-1} (21). Thus, the first nitrogen derivative of Cu(hfac)$_2$ that we studied was Cu(hfac)$_2$·2N$_2$H$_4$, which was originally reported by Bublik and co-workers (33). We find that this adduct can be used as a Cu CVD precursor, with Cu film deposition occurring at substrate temperatures as low as 140 °C under N$_2$.

$$H_2N-NH_2 \longrightarrow N_2 + 2H_2 \qquad (12)$$

Substituted hydrazines are not as strongly reducing as the parent N$_2$H$_4$. However, a number of 1,2-disubstituted hydrazines can be oxidized to azo compounds (diazenes). For example, ΔH for the dehydrogenation of 1,2-dimethylhydrazine to azomethane (CH$_3$N=NCH$_3$; see reaction 13 above) can be estimated to be –60 kJ/mol (30) (34). The starting compounds RNH–NHR also have the advantage that they are less hazardous than N$_2$H$_4$, both to safety and health. Accordingly, we are now preparing the Cu(hfac)$_2$ adducts with CH$_3$NH–NHCH$_3$ and related hydrazines. Preliminary experiments show that these complexes yield Cu on heating under N$_2$.

$$RCH_2-NHR' \longrightarrow RCH=NR' + H_2 \qquad (13)$$

We are now comparing these hydrazine derivatives to other amine adducts. We have recently prepared Cu(hfac)$_2$·piperidine; see sketch at right and crystal-structure drawing in Figure 1. This adduct melts at 84-85 °C, and is thus a liquid under typical CVD evaporator conditions. (It also has an unusual type of square-pyramidal structure, in which the neutral ligand occupies an *equatorial* site, forcing one of the hfac O atoms into the apical site.) We have

successfully deposited Cu films using this precursor at substrate temperatures of 230 °C under H$_2$. In contrast, experiments with N$_2$ as carrier gas (at temperatures to 260 °C) have not been successful. Thus, as suggested by the thermodynamic data cited above, it appears to be more difficult to use simple amines as reducing agents in place of H_2.

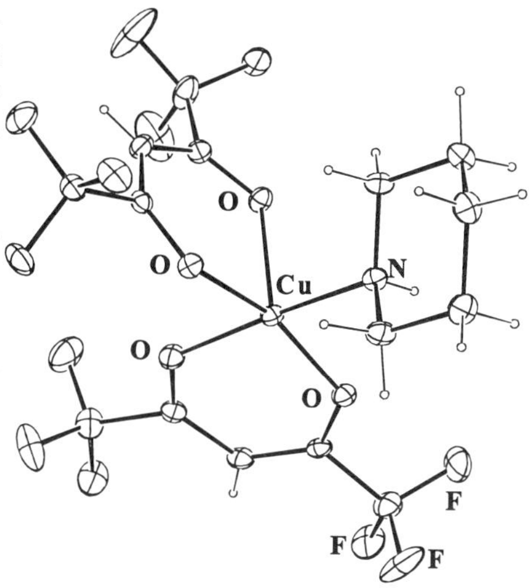

New copper(I) precursors. We were interested in copper(I) CVD precursors outside the (hfac)Cu^I family, for two reasons. First, a Cu(I) precursor containing no O or F atoms may be useful for co-deposition of Cu–Al alloys. Cu containing small amounts of Al is substantially harder than pure Cu, with resistivity nearly as low as that of the pure metal. However, if the two metals are to be deposited simultaneously, the precursors

Figure 1. Structure of Cu(hfac)$_2$·piperidine.

must be compatible with each other as well as with the two metals. Since Al and its common CVD precursors (e.g. AlEt$_3$, Et$_2$AlH, AlH$_3$·NR$_3$) can react with O and F compounds, they may be incompatible with Cu–hfac complexes. One successful example of co-deposition has been reported by Kondoh et al., who used Et$_2$AlH and CpCuPEt$_3$ as their Al and Cu precursors (35).

Our second reason for pursuing new copper(I) precursors is to prepare volatile species with readily accessible excited states for photoinduced deposition reactions. Cu(I) complexes have been extensively studied by photochemical methods (36); however, most of the photoactive species described so far have been involatile (many because they are ionic salts). In a photochemical deposition process, for example, reaction from an excited state of the precursor is more favorable than from its ground state; thus, the photochemical process can occur faster or at a lower temperature than the ground-state reaction. Potential applications of photoinduced deposition processes have been reviewed by Ibbs & Osgood (37). One of the more promising is to use a photochemical process to prepare a thin layer of Cu, followed by deposition of a thicker film by conventional CVD; the photodeposited "seed layer" may enhance the selectivity of the overall deposition process. We have examined two Cu(I) species as potential CVD precursors: the borohydride complex (2,9-Me$_2$phen)CuBH$_4$, and the tetrameric amide cluster [CuN(SiMe$_3$)$_2$]$_4$.

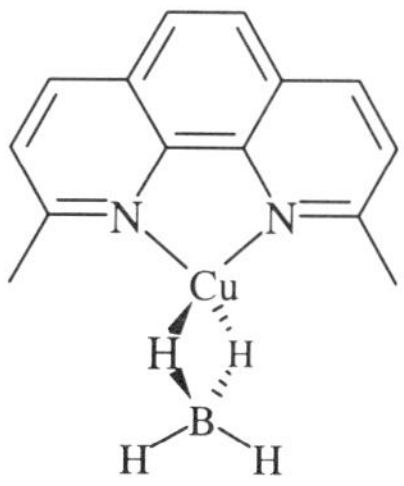

We envisioned that (2,9-Me$_2$phen)CuBH$_4$ (see sketch at right) would have two attractive features. First, its coordinated 2,9-dimethylphenanthroline ligand should provide a well-defined framework for metal-to-ligand charge-transfer (MLCT) excited states. Many copper(I) complexes with polypyridine ligands show intense absorption in the 400-500 nm range, and phosphorescence near 600 nm. The second important aspect of the complex is its borohydride ligand. This makes the overall complex

neutral, as opposed to the more common Cu(polypyridine)$_n^+$ species; also, the coordinated BH$_4$ ligand may act as a reducing agent for the Cu, removing the need for a reducing carrier gas. Unfortunately, although the complex is luminescent (see phosphorescence spectrum in Figure 2), it is not sufficiently volatile for CVD experiments.

For our other experiments in this area, we have concentrated on the tetrameric copper(I) amide cluster [CuN(SiMe$_3$)$_2$]$_4$ (38) (see sketch below at right). This complex is colorless, absorbing intensely in the

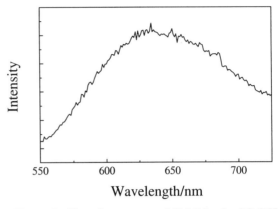

Figure 2. Phosphorescence of (2,9-Me$_2$phen)CuBH$_4$.

UV region (λ_{max} 283, 246 nm), and it shows intense blue-green luminescence (λ_{max} ca. 510 nm) under a variety of conditions (39). In contrast to the MLCT states accessible for Cu(I) polypyridine complexes, the low-lying excited states of this amide cluster are likely to be weakly Cu–Cu bonding in nature.

[CuN(SiMe$_3$)$_2$]$_4$ is less volatile than typical Cu^{II}(hfac)$_2$ and Cu^{I}(hfac)L complexes (vapor pressure ca. 0.2 Torr at 180 °C (38)), but it can still be used as a CVD precursor. The temperatures required for evaporation and Cu metal formation with this precursor are similar. Therefore, in our experiments, the solid precursor was placed on or immediately upstream of the substrate. Small amounts of Cu film deposition were observed at substrate/evaporation temperatures as low as 145 °C (1 atm H$_2$, 3 h, Si substrates). At 200 °C, deposition rates were higher (see Table IV), but still well below those for the Cu(hfac)$_2$ adducts.

We also studied deposition using this precursor under UV illumination (Xe arc lamp). Under these conditions Cu deposition occurred at a slightly lower minimum temperature (136-138 °C), suggesting a slight photochemical enhancement of the deposition process. Care was taken during these experiments to monitor the actual temperature of the substrate surface being irradiated, so that the measured temperature included any heating of the substrate by the arc lamp beam.

The above results suggest that copper(I) amide complexes may be suitable precursors for both conventional and photoinduced CVD processes. However, both the deposition rate for the non-photochemical process and the degree of enhancement under illumination need to be improved. Thus, we have attempted to make more volatile Cu(I) complexes that still retain the Cu–NR$_2$ "functional group"; our principal target has been complexes of the form L$_n$CuN(SiMe$_3$)$_2$ (L = phosphine, alkene).

Our normal method for preparing [CuN(SiMe$_3$)$_2$]$_4$, from CuCl and LiN(SiMe$_3$)$_2$, still gives [CuN(SiMe$_3$)$_2$]$_4$ as the main isolated product even when the reaction is carried out in the presence of an excess of the phosphines PEt_3, Ph_3P, $Me_2PCH_2CH_2PMe_2$ (dmpe), or $Ph_2PCH_2CH_2PPh_2$ (dppe). However, preliminary experiments with $(CH_3CN)_4Cu^+$ as starting complex have been more promising. Treatment of $(CH_3CN)_4Cu^+$ first with dmpe or dppe, followed by $LiN(SiMe_3)_2$, yields yellow materials whose luminescence (λ_{max} ca. 550 nm) is clearly different from that of $[CuN(SiMe_3)_2]_4$. These products are likely to have monomeric or dimeric structures, as shown in the sketch above at right; structures with bridging $N(SiMe_3)_2$ ligands are less likely, because they would require higher coordination numbers at Cu (40). The product with dmpe is of special interest because it produces Cu films on heating. Conventional and light-induced CVD experiments with this precursor are now in progress.

Summary

We have discussed recent progress with three families of Cu CVD precursors. *(a) $Cu^{II}(hfac)_2$–alcohol adducts:* Alcohols can be efficient reducing agents for $Cu(hfac)_2$, permitting Cu deposition rates with N_2 carrier gas nearly three times as great as with H_2 under otherwise identical conditions. The best results thus far have been obtained with the isopropyl alcohol adduct. *(b) Adducts of $Cu^{II}(hfac)_2$ with nitrogen bases:* $Cu(hfac)_2 \cdot L$ (e.g. L = piperidine) can be used for Cu CVD with H_2 carrier gas. The adduct with hydrazine, on the other hand, is self-reducing, leading to Cu film formation under N_2 at substrate temperatures of 140 °C. *(c) Photosensitive copper(I) precursors:* The phosphorescent amide cluster $[CuN(SiMe_3)_2]_4$ can be used for Cu CVD under H_2. Under UV irradiation, it shows a slight lowering of the threshold substrate temperature for Cu deposition; this suggests that the deposition can be accelerated photochemically.

Experimental Section

Anhydrous $Cu(hfac)_2$ was prepared from commercial $Cu(hfac)_2 \cdot H_2O$ (Gelest or Strem) by dehydration over H_2SO_4 or P_2O_5 (8), followed by vacuum treatment to remove the last traces of water. Adducts of $Cu(hfac)_2$ with amines were prepared by treating the anhydrous compound with a stoichiometric quantity of the base in CH_2Cl_2 solution, followed by removal of solvent and recrystallization or sublimation. Several of the amines were most readily available as hydrochloride salts (e.g. N_2H_4). In these cases, the hydrochloride was suspended in CH_2Cl_2 in the presence of excess NaOH(s), with stirring, and anhydrous $Cu(hfac)_2$ added when the reaction with NaOH appeared to be complete; the resulting suspension was filtered and evaporated to yield the product. The N_2H_4 adduct occasionally decomposed during this process. (*Caution!* We did not observe any violent decomposition in this system, but it should be handled with care, because hydrazine is potentially explosive and highly toxic.) Alcohol adducts were prepared by the general method of Baum et al. (14): $Cu(hfac)_2$ was dissolved in an excess of the appropriate alcohol in a drybox, followed by removal of solvent and recrystallization or sublimation. Other precursors were prepared by literature procedures.

CVD experiments were carried out in a Pyrex warm-wall test reactor which is similar to a reactor we have described previously (*3*), except that its susceptor is 2.5 cm in diameter. Substrates were borosilicate glass or Si (with native oxide). Resistivity (Veeco four-point probe) and thickness (Tencor stylus profilometer) measurements were performed in five places on each film; the values for a single film normally agreed within ±20%. The purity of the films deposited using $Cu(hfac)_2 \cdot i\text{-PrOH}$ under N_2 and H_2 was identical (based on Auger electron spectra, within experimental error) to that of Cu foil (99.9%). Thin Cu films prepared from some of the other precursors (e.g. $[CuN(SiMe_3)_2]_4$) were electrically discontinuous; they were identified as Cu metal, and their thicknesses estimated, by powder X-ray diffraction.

In photochemical experiments, a 150-W Xe arc lamp was used, which provided a light intensity of ca. 10^{16} photons/s over the irradiated substrate (300-400 nm), or ca. 10^{20} photons during a typical 3-h deposition experiment.

Acknowledgments. We are grateful to Dr. Frank R. Fronczek for the X-ray structure analyses discussed here, and to Dr. Gregory L. Griffin for numerous helpful discussions. This research was supported by the National Science Foundation (CTS-9612157) and the Department of Energy (EPSCoR). A. M. J. thanks the Louisiana Board of Regents for a graduate fellowship.

Literature Cited

1. (a) Singer, P. *Semicond. Intl.* **1997**, *20* (No. 13, November), 67-70. (b) "IBM introduces advanced chip technology", IBM Corp., East Fishkill, NY, September 1997 (http://www.chips.ibm.com/news/cmos7s.html).

2. Kodas, T. T.; Hampden-Smith, M. J., Eds. *Chemistry of Metal CVD*; VCH: Weinheim, 1994; chapters 4 and 5.

3. Kumar, R.; Fronczek, F. R.; Maverick, A. W.; Lai, W. G.; Griffin, G. L. *Chem. Mater.* **1992**, *4*, 577-582

4. Reynolds, S. K.; Smart, C. J.; Baran, E. F.; Baum, T. H.; Larson, C. E.; Brock, P. J. *Appl. Phys. Lett.* **1991**, *59*, 2332-2334. Jain, A.; Chi, K.-M.; Hampden-Smith, M. J.; Kodas, T. T.; Farr, J. D.; Paffett, M. F. *J. Mater. Res.* **1992**, *7*, 261-264.

5. Doppelt, P.; Baum, T. H.; Ricard, L. *Inorg. Chem.* **1996**, *35*, 1286-1291; and references therein.

6. Norman, J. A. T.; Muratore, B. A.; Dyer, P. N.; Roberts, D. A.; Hochberg, A. K. *J. Phys. (Paris) IV*, **1991**, *1 (Coll. C2)*, 271. Norman, J. A. T.; Muratore, B. A.; Dyer, P. N.; Roberts, D. A.; Hochberg, A. K.; Dubois, L. H. *Mater. Sci. Eng.* **1993**, *B17*, 87-92. Jain, A.; Chi, K.-M.; Kodas, T. T.; Hampden-Smith, M. J. *J. Electrochem. Soc.* **1993**, *140*, 1434-1439.

7. Nguyen, T.; Charneski, L. J.; Hsu, S. T. *J. Electrochem. Soc.* **1997**, *144*, 2829-2833.

8. Funck, L. L.; Ortolano, T. R. *Inorg. Chem.* **1968**, *7*, 567-573.

9. Walker, W. R.; Li, N. C. *J. Inorg. Nucl. Chem.* **1965**, *27*, 2255-2261.

10. Partenheimer, W.; Drago, R. S. *Inorg. Chem.* **1970**, *9*, 47-52.

11. Nozari, M. S.; Drago, R. S. *Inorg. Chem.* **1972**, *11*, 280-283. McMillin, D. R.; Drago, R. S.; Nusz, J. A. *J. Am. Chem. Soc.* **1976**, *98*, 3120-3126.

12. Gafney, H. D.; Lintvedt, R. L. *J. Am. Chem. Soc.* **1971**, *93*, 1623-1628.

13. Houle, F. A.; Wilson, R. J.; Baum, T. H. *J. Vac. Sci. Technol. A* **1986**, *4*, 2452-2458.

14. (a) Kovac, C. A.; Jones, C. R.; Baum, T. H.; Houle, F. A. *IBM Res. Rep. No. RJ 4174 (46102),* **1984**. (b) Doppelt, P.; Baum, T. H. *Mater. Res. Soc. Bull.* **1994**, *19*(8), 41-48.

15. Awaya, N.; Arita, Y. In: *Advanced Metalization for ULSI Applications in 1991*, Rana, V. V. S.; Joshi, R. V.; Ohdomari, I., Eds.; Materials Research Society: Pittsburgh, PA, 1992; p. 345. Awaya, N.; Arita, N. *Jpn. J. Appl. Phys. (1)* **1993**, *32*, 3915-3919.

16. Cho, C.-C. In: *Tungsten and Other Advanced Metals for ULSI Applications in 1990*; Smith, G. C.; Blumenthal, R., Eds.; Materials Research Society: Pittsburgh, PA, 1991; p. 189. Cho, C.-C. U. S. Patent 5,087,485 (1992).

17. Zheng, B.; Eisenbraun, E. T.; Liu, J.; Kaloyeros, A. E. *Appl. Phys. Lett.* **1992**, *61*, 2175.

18. Pilkington, R. D.; Jones, P. A.; Ahmed, W.; Tomlinson, R. D.; Hill, A. E.; Smith, J. J.; Nuttall, R. *J. Phys. (Paris) IV*, **1991**, *1 (Coll. C2)*, 263-269.

19. (a) Lai, W. G.; Xie, Y.; Griffin, G. L. *J. Electrochem. Soc.* **1991**, *138*, 3499-3504. (b) Wang, J.; Little, R. B.; Lai, W. G.; Griffin, G. L. *Thin Solid Films* **1995**, *262*, 31-38. (c) Borgharkar, N. S.; Griffin, G. L. *J. Electrochem. Soc.* **1998**, *145*, 347-352. The mechanism presented by these authors, shown here in somewhat simplified form, also included the inhibiting effect of added hfacH on the reaction.

20. (a) Dissociative adsorption of alcohols on Cu: Bowker, M.; Madix, R. J. *Surf. Sci.* **1980**, *95*, 190-206. Russell, J. N., Jr.; Gates, S. M.; Yates, J. T., Jr. *Surf. Sci.* **1985**, *163*, 516-540. (b) Dehydrogenation of *i*-PrOH over Cu: Cunningham, J.; Al-Sayyed, G. H.; Cronin, J. A.; Fierro, J. L. G.; Healy, C.; Hirschwald, W.; Ilyas, M.; Tobin, J. P. *J. Catal.* **1986**, *102*, 160-171. (c) Generation of carbonyl compounds from alcohols in plasmas: Kaloyeros, A. E.; Zheng, B.; Lou, I.; Lau, J.; Hellgeth, J. W. *Thin Solid Films*, **1995**, *262*, 20-30.

21. Afeefy, H. Y.; Liebman, J. F.; Stein, S. E. "Neutral Thermochemical Data"; In: *NIST Standard Reference Database Number 69*; Mallard, W. G.; Linstrom, P. J., Eds. August 1997, National Institute of Standards and Technology, Gaithersburg MD 20899 (http://webbook.nist.gov).

22. Lide, D. R., Ed.; *CRC Handbook of Chemistry and Physics*; 77th Ed.; CRC Press: Boca Raton, FL, 1996; and references therein. Some values in these tables are based on enthalpies of combustion and vaporization in: Weast, R. C., Ed.; *CRC Handbook of Chemistry and Physics*; 62nd Ed.; CRC Press: Boca Raton, FL, 1981.

23. Serjeant, E. P.; Dempsey, B. *Ionisation Constants of Organic Acids in Aqueous Solution*; Pergamon: London, 1979.

112

24. King, E. J. "Acid-base behaviour"; In: *Physical Chemistry of Organic Solvent Systems*; Covington, A. K; Dickinson, T., Eds.; Plenum: London, 1973; Ch. 3.

25. Hunter, E. P.; Lias, S. G. "Proton Affinity Evaluation"; In: *NIST Standard Reference Database Number 69*; Mallard, W. G.; Linstrom, P. J., Eds. August 1997, National Institute of Standards and Technology, Gaithersburg MD 20899 (http://webbook.nist.gov).

26. Griffin, G. L.; Borgharkar, N. S.; Maverick, A. W.; Fan, H. In: *Advanced Metalization and Interconnect Systems for ULSI Applications in 1995*; Ellwanger, R.; Wang, S.-Q., Eds. Materials Research Society: Pittsburgh, PA, 1996; pp 195-199.

27. Chiang, C.-M.; Miller, T. M.; Dubois, L. H. *J. Phys. Chem.* **1993**, *97*, 11781-11786.

28. Pinkas, J.; Huffman, J. C.; Chisholm, M. H.; Caulton, K. G. *Inorg. Chem.* **1995**, *34*, 5314-5318.

29. Pinkas, J.; Huffman, J. C.; Bollinger, J. C.; Streib, W. E.; Baxter, D. V.; Chisholm, M. H.; Caulton, K. G. *Inorg. Chem.* **1997**, *36*, 2930-2937.

30. Pedley, J. B.; Naylor, R. D.; Kirby, S. P. *Thermochemical Data of Organic Compounds*; Chapman and Hall: London, 1986.

31. Wiberg, K. B.; Nakaji, D. Y.; Morgan, K. M. *J. Am. Chem. Soc.* **1993**, *115*, 3527-3532.

32. Häfelinger, G.; Steinmann, L. *Angew. Chem. Int. Ed. Engl.* **1977**, *16*, 47-48. Jackman, L. M.; Packham, D. I. *Proc. Chem. Soc.* **1957**, 349-350.

33. Bublik, Zh. N.; Volkov, S. V.; Mazurenko, E. A. *Russ. J. Inorg. Chem.* **1984**, *29*, 73-76.

34. Rossini, F. D.; Montgomery, R. L. *J. Chem. Thermodyn.* **1978**, *10*, 465-470.

35. Kondoh, E. Kawano, Y.; Takeyasu, N.; Ohta, T. *J. Electrochem. Soc.* **1994**, *141*, 3494-3499.

36. Kutal, C. *Coord. Chem. Rev.* **1990**, *99*, 213-252.

37. Ibbs, K. G.; Osgood, R. M., Eds. *Laser Chemical Processing for Microelectronics*; Cambridge University Press: Cambridge, U. K., 1989.

38. Bürger, H.; Wannagat, U. *Monatsh. Chem.* **1964,** *95*, 1099-1102.

39. James, A. M.; Laxman, R. K.; Fronczek, F. R.; Maverick, A. W. *Inorg. Chem.* in press.

40. A side product in the reaction of $(CH_3CN)_4Cu^+$ with dppe is the binuclear complex $(dppe)[Cu(NCCH_3)_2]_2^{2+}$, whose structure we have determined by X-ray analysis (James, A. M.; Fronczek, F. R.; Maverick, A. W. unpublished work). This complex contains trigonal planar Cu(I) centers with weak Cu-Cu interactions (Cu⋯Cu 2.872(1) Å). (Electronically similar d^{10}-d^{10} interactions have been described in dimeric Pd(0) and Pt(0) complexes: Caspar, J. V. *J. Am. Chem. Soc.* **1985**, *107*, 6718-6719. Harvey, P. D.; Gray, H. B. *J. Am. Chem. Soc.* **1988**, *110*, 2145-2147. Maguire, N. A. P.; Wright, L. L.; Guckert, J. A.; Tweet, W. S. *Inorg. Chem.* **1988**, *27*, 2905-2907.) This suggests that a dimeric structure is reasonable for the $[(R_2PCH_2CH_2PR_2)CuN(SiMe_3)_2]_n$ species.

Chapter 9

New Materials for Si-Based Heterostructure Engineering: Synthesis of Si–Ge–C Alloys and Compounds by UHV-CVD and Molecular Chemistry

J. Kouvetakis[1], D. C. Nesting[1], and David J. Smith[2]

[1]Department of Chemistry and Biochemistry and [2]Center for Solid State Science and Department of Physics, Arizona State University, Tempe, AZ 85287

Inorganic precursors have been synthesized and used to deposit novel group IV heterostructures. Metastable SiGeC alloys offer the prospect of strain compensation in the pseudomorphic SiGe system, as well as providing the possibility of changes in the bandgap energy to values greater than those of Si and SiGe. We have demonstrated the use of the C-H free, $C(SiH_3)_4$ and $C(GeH_3)_4$, precursors to grow pseudomorphic $(Si_4C)_xGe_y$, $(Ge_4C)_xSi_y$, and $Si_{1-x-y}Ge_xC_y$ alloys containing 4-7 at. % C the highest C content incorporated in crystalline SiGeC. The materials are deposited on Si by ultrahigh vacuum (UHV) CVD and characterized for structure, composition and bonding properties by RBS, high-resolution TEM, electron energy loss spectroscopy and FTIR and Raman spectroscopies, respectively. Non-stoichiometric GeC alloys are also potential candidates for bandgap engineering and lattice matching with Si. The synthetic routes involve UHV-CVD reactions of GeH_4 with $HC(GeH_3)_3$ that produce $Ge_{1-x}C_x$ alloys with up to 7 at. % C. The incorporation of C into the Ge lattice completely suppresses island like-film formation and results in 2-D layer-by-layer perfectly epitaxial growth. Synthesis of new ordered materials composed of C, Si, and Ge that incorporate carbon on a common sublattice is also described. Examples of these include new diamond-structured materials with composition Si_4C, and $(Si_2Ge)C_x$ as well as a new sphalerite phase with composition $GeSi_3C_4$. Finally we present synthesis and structural analysis of the $CH(GeH_3)_3$ and $C(GeH_3)_4$ precursors utilized in these studies, as well as synthesis of corresponding halide cluster compounds $C(GeBrX_2)_4$ (X=Cl, Br).

Introduction

Modern device design and fabrication is intimately associated with the concept of bandgap engineering. This concept has been applied to realize practical goals such as the creation of light-emitting and laser diodes, faster transistors, and a series of other novel devices using group III-V semiconductors. However, the spectacular advances in integrated microelectronics over the past 35 years have primarily relied on the unique properties of Si. Industrial ingenuity combined with key technological breakthroughs have made it possible to double the transistor density in electronic devices about every three years. Thus, the synthesis of bandgap engineered, heterostructures on Si should be important for future generations of high-speed devices. Research and development in this area has been focused on the Si-Ge system but there has been increasing interest and activity in other group IV combinations.

The Si-Ge system. The complete miscibility of Si and Ge allows growth of $Si_{1-x}Ge_x$ alloys with variable compositions and tailored band structure. The addition of the slightly larger Ge in Si produces a lattice constant larger than that of Si (a_{Si} =5.43Å, a_{Ge}=5.65Å). For this reason, $Si_{1-x}Ge_x$ layers grown heteroepitaxially on Si are under an inherent compressive strain. This strain combined with variable compositions result in layers with smaller gaps than that of Si. Typically, the bandgap decreases monotonically from 1.12 eV for pure Si to 0.67 eV for pure Ge as the Ge composition in the alloy increases. The bandgap reduction has important advantages in device applications. Some of these include higher emitter injection efficiency, lower base resistance, faster switching times, and improved performance at lower temperatures.

As a consequence of these properties, a variety of high performance electronic devices such as heterojunction bipolar transistors, heterojunction field effect transistors and photodetectors based on $Si_{1-x}Ge_x$ have been created and successfully integrated with Si technology (1-3). A major drawback in the use of $Si_{1-x}Ge_x$ heterostructures is that growth of perfectly epitaxial films, essential for effective device performance, is hindered by the small lattice mismatch between SiGe and Si. Epitaxial layers of SiGe can only be grown by adopting the smaller lattice constant of Si. This adjustment in lattice constant, known as strained layer epitaxy, causes compressive strain in the layer, thus making the epitaxial layer unstable with increasing thickness and Ge content. The maximum thickness of a strained layer before it becomes unstable is termed critical thickness. Upon exceeding the critical thickness, the material relieves strain by injecting misfit dislocations at the interface as illustrated in Figure 1. These defects are detrimental to the device performance.

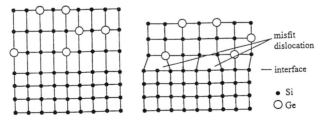

Figure 1   Schematic of a $Si_{1-x}Ge_x$ heterostructure;   a) commensurate tetragonally-distorted layer under strain; b) relaxed layer unstrained

Carbon incorporation: Development of metastable diamond-like systems. It has been proposed that addition of a third element with smaller size such as C, B, or P could reduce strain by forming a ternary alloy that is lattice matched with the Si substrate. C is particularly attractive because it forms four equivalent Si-C bonds that are much shorter than the normal Si-Si bonds. It has been known for decades that C is a common impurity in Si and it exists as substitutional C of T_d symmetry which is normally identified by IR as a single local mode with a frequency of 607 cm^{-1}(4). Although C can exist as an isolated impurity in the Si lattice, the solubility of C in Si is negligible, 10^{-6} at.% at the melting point, and non-existent in the analogous Ge.

Carbon incorporation onto substitutional sites in Si, Ge, SiGe, or even Sn, as a potential method for tailoring the structural, and electronic properties of these materials has become the focus of considerable research. The current advances in SiGe technology have prompted efforts to develop other crystalline group IV binary and ternary alloys, an area of research which had remained virtually unexplored until very recently. It is believed that a major limitation for formation of group IV alloys and compounds is the large difference in covalent radii among the elements, ranging from 0.77 Å for C to 1.40 Å for Sn. Another limitation is that the diamond structure is stable for Si and Ge, metastable for C, and only stable up to almost room temperature for a-Sn. For these reasons, most alloys and ordered phases involving combinations of C, Si, Ge, and Sn are metastable and the preparation of such materials will only be possible under conditions that are far from equilibrium. Only SiC with its numerous polytypes and Si-Ge are thermodynamically stable systems that have been thoroughly studied. Despite this metastability, intense efforts are under way to produce materials that combine the unique and exciting properties that the group IV elements are known to possess. These include the highest thermal conductivity (diamond), high hole mobility (Ge), bandgap variation from 5.5 to 0 eV and widespread technological applications (Si). The major challenge for group IV alloy or compound formation is development of novel synthetic methods that yield device quality materials under metastable conditions.

Although bandgap engineering and lattice matching with Si are the major issues that drive group IV alloy research, the quest for a direct bandgap semiconductor that would integrate microelectronics and optoelectronics has yet to be satisfied. Present day microelectronics is based primarily on Si technology with several million transistors per chip, but this technology cannot be applied to optoelectronics because of the indirect bandgap of group IV semiconductors. The direct energy gap for Ge lies slightly above the indirect gap, and a transition to a direct band gap has been predicted for metastable SnGe alloys. The prospect for creating a direct band gap material in the Si-Ge-C, Ge-C or Sn-Ge-C systems is uncertain but potentially ground breaking.

$Si_{1-x-y}Ge_xC_y$ and $Ge_{1-x}C_x$. The successful creation of bandgap engineered heterojunction bipolar transistors using $Si_{1-x}Ge_x$ alloys has led to a tremendous interest in $Si_{1-x-y}Ge_xC_y$ alloys. The motivation of this work is to develop an epitaxial bandgap engineerable material that lattice matches Si and is compatible with group IV ULSI technology. Incorporation of C in sufficient amounts into SiGe alloys may result in an increase of the bandgap to values greater than those of SiGe and Si. Theoretical investigations based on an empirical interpolation technique proposed that the bandgap of $Si_{1-x-y}Ge_xC_y$ should increase with increasing C concentration (5). The estimated band gaps span the entire 0.62-5.5 eV range as shown in Figure 2.

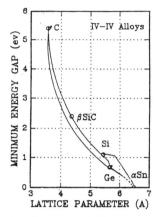

Figure 2   Indirect energy band gap vs. lattice parameter of group IV-IV heterostructures on Si.

Other theoretical studies of the $Si_{1-x}C_x$ system predict that the bandgap of the alloy at low C concentrations is smaller than the bandgap of pure Si (6). Preliminary experimental studies of Si/SiGeC capacitors in which the $Si_{1-x-y}Ge_xC_y$ (y=0.02) heterostructure is compressively strained show that the valence band offset decreases with increasing C and that the conduction band offsets are negligible (7). Alternatively, substitutional C in Ge-rich $Si_{1-x-y}Ge_xC_y$ has been found to increase the bandgap (8). These contradicting results have stimulated considerable experimental interest in this intriguing material.   In addition to bandgap engineering, the incorporation of C is also expected to reduce the lattice mismatch between $Si_{1-x}Ge_x$ and Si, with the smaller size of C compensating for the larger size of Ge.   Diamond has a lattice constant of 3.545 Å, significantly smaller than those of Si (5.43 Å) and Ge (5.646 Å). In theory, C, Si, and Ge can be intermixed to form metastable diamond-like alloys that should have lattice constants smaller than those of the $Si_{1-x}Ge_x$ alloys.

Although SiGeC compositions with bandgaps greater than that of Si have not yet been developed, it has been demonstrated that even modest concentrations of the mismatched elements can have substantial structural impact(Figure3). For example, high quality epitaxial layers of $Si_{1-x-y}Ge_xC_y$ (y=0.02) alloys have been grown on Si with thickness substantially greater than the critical thickness of pure $Si_{1-x}Ge_x$ on Si

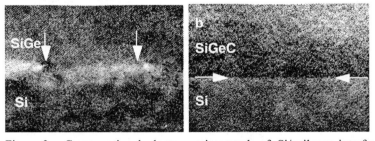

Figure 3.   Cross-sectional electron micrograph of Si/epilayer interface: a) Si/SiGe interface with misfit dislocations (arrowed) b) Si/Si-Ge-C interface (arrowed) showing defect-free coherent growth.

Substitution of C in the SiGe lattice has reduced the lattice mismatch between SiGe and the Si substrate. If Vegards' Law applies to this system, a 9:1 Ge to C ratio is required to produce Si lattice matched-alloys. It has proven difficult, however, to incorporate C in sufficient quantities to achieve lattice matching, especially for large Ge concentrations. The major problems are SiC precipitation and defect formation in highly strained materials.

Methods to deposit epitaxial $Si_{1-x-y}Ge_xC_y$ films were based on molecular beam epitaxy (MBE) and solid phase epitaxy (SPE) (9-10) and gave only 1-2 at. % C and a limited range of Si:Ge compositions. Recently, CVD processes using conventional hydrocarbons with multiple C-H bonds (C_2H_2, C_2H_6) as the source of C have been utilized to grow device quality material with 2 at. % C incorporation. The addition of C had a marked effect on the film growth. Whereas misfit dislocations were usually observed at the SiGe/Si interface due to the lattice mismatch, these SiGeC films had very few interfacial defects. Furthermore, it was observed that the addition of C into the SiGeC lattice led to substantial increases in the critical thickness for defect-free growth. Despite the successful synthesis of device quality, low C content SiGeC films, continuing efforts to introduce greater than 2 at.%C required high deposition temperatures and have usually resulted in amorphous or at best polycrystalline materials. The low C incorporation is attributed to the stability of the C-H and C-C bonds of the precursors which require high deposition temperatures that favor carbide precipitation, phase separation, and inhomogeneous growth.

It is now generally accepted that high C concentrations are necessary to obtain $Si_{1-x-y}Ge_xC_y$ materials with the desired lattice constants and band offsets. In order to study the influence of composition on the bandgap and the lattice constant, we are developing new methods that provide a higher degree of compositional control and allow the lower deposition temperatures that are considered necessary to prepare metastable structures in the Si-Ge-C system. One of our long-term objectives is to explore alternative ways of creating highly concentrated layers (C>4 at. %) that are inaccessible by conventional routes. We emphasize an integrated approach involving novel precursor chemistries and low temperature (UHV CVD), an inherently non-equilibrium process, to prepare these thermodynamically metastable materials.

Results

$C(SiH_3)_4$ and $C(SiH_3)_4$: New methods for Si-Ge-C growth. We have succeeded in growing heteroepitaxial $Si_{1-x-y}Ge_xC_y$ alloys containing up to 6 at. % substitutional C which is the highest content observed in crystalline SiGeC alloys (11-12). Film growth was achieved by reactions of $C(GeH_3)_4$, and $C(SiH_3)_4$, with SiH_4 and GeH_4 on atomically clean (100) Si surfaces. $C(GeH_3)_4$ and $C(SiH_3)_4$ were chosen for deposition for several reasons: a) The molecular arrangement of a single carbon atom that occupies a diamond-like site surrounded only by four Si or Ge atoms is structurally consistent with the desired material and the Si substrate; b) The encapsulation of C between four SiH_3 or GeH_3 ligands simplifies the formidable task of forming SiGeC by reaction of hydrocarbons with Si and Ge hydrides to a Si-hydride Ge-hydride interaction; c) The absence of strong C-H bonds should favor low-temperature growth leading to metastable materials. d) $C(SiH_3)_4$ is a volatile solid (70 Torr, 22°C) and, unlike other Si hydrides, it appears to be non-toxic and non-pyrophoric. $C(GeH_3)_4$ is also a volatile (2 Torr, 22°C) waxy solid and decomposes very slowly in air. e) C(SiH$_3$)$_4$ is thermally robust at the growth conditions. It decomposes only at temperatures higher than 600°C via loss of H$_2$ to yield Si-C films with composition Si$_{80}$C$_{20}$. C(GeH$_3$)$_4$ is thermally stable below 500 °C and it decomposes readily at 530 °C to yield films with composition Ge$_{82}$C$_{18}$. We prepared C(SiH$_3$)$_4$ in high purity by a procedure described by Schmidbaur (14). C(GeH$_3$)$_4$ was prepared for the first time using a procedure described later.

Synthesis and characterization of (Si$_4$C)$_x$Ge$_y$ and (Ge$_4$C)$_x$Si$_y$. We investigated first reactions of C(SiH$_3$)$_4$ with GeH$_4$ and C(GeH$_3$)$_4$ with SiH$_4$ to produce materials with the general formulas (Si$_4$C)$_x$Ge$_y$ and (Ge$_4$C)$_x$Si$_y$ respectively. In a typical deposition, we find that C(SiH$_3$)$_4$ reacts readily with GeH$_4$ at 450°C to yield SiGeC materials that have Si to C compositions in the same ratio 4:1 as the precursor. We prepared a range of (Si$_4$C)$_x$Ge$_y$ compositions containing 3-10 at. % C. TEM examination revealed that a typical sample having C=5 at. %, Si= 20 at. %, Ge=75 at. % was monocrystalline and epitaxial. We observed, however, that severe degradation in the epitaxial quality occurred with the C content increased to levels above 7 at. %. Samples with 10 at.% C were entirely polycrystalline. TEM examinations of (Ge$_4$C)$_x$Si$_y$ materials revealed that a typical sample with C=5 at. %, Ge=20 at. % and Si=75 at. % was completely epitaxial and displayed better crystallinity and fewer defects than the sample with the same C concentration obtained by using C(SiH$_3$)$_4$ (Figure 4). The reactions of C(SiH$_3$)$_4$ and C(GeH$_3$)$_4$ with GeH$_4$ and SiH$_4$ respectively demonstrate preparation of single phase material incorporating a significant C content (5-6 %). They also indicate the remarkable degree of compositional control that these precursors provide by incorporating the composition of the Ge$_4$C and Si$_4$C molecular framework into the solid state.

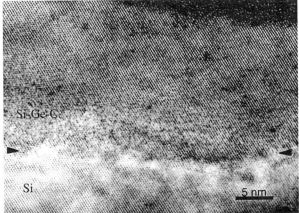

Figure 4. Cross sectional electron micrograph showing hetroepitaxial growth of (Ge$_4$C)$_x$C$_y$ (y=0.05) obtained from reactions using C(GeH$_3$)$_4$ as the carbon source. Note relatively defect-free Si/SiGeC interface.

In addition to high-resolution TEM that provided information about local microstructure, the other characterization techniques that were used routinely to determine structural compositional and bonding properties of the materials were: a)

Rutherford backscattering spectroscopy (RBS) to establish elemental composition, estimate film thickness and determine quality of epitaxial growth by channeling; b) Secondary ion mass spectrometry (SIMS) for quantitative depth profiling and identification of the surface elements; c) transmission FTIR and high-resolution Raman to identify the local bonding environment of the C in the lattice, the nature of possible precipitates, and to investigate strain; c) Electron-energy-loss spectroscopy (EELS) to obtain hybridization of C and examine elemental distribution at the nanometer scale.

Synthesis of $Si_{1-x-y}Ge_xC_y$: We also investigated reactions of $C(SiH_3)_4$ at 470^oC with combinations of SiH_4 and GeH_4 to produced $Si_{1-x-y}Ge_xC_y$ films with as 4-7 at. % C. The aim of these experiments was to explore preparation of materials that incorporated wide variations in Si and Ge while maintaining the C content in the 5% range. With 6 Å per min. growth rates, films of $Si_{1-x-y}Ge_xC_y$ (x=0.31-0.35 and y=0.04-0.07) were grown to produce 1100 Å layers that were nominally lattice-matched to Si. TEM images of the Si-SiGeC interface for a C content of 4 at. %, demonstrated that a high degree of crystalline perfection was obtained across the interface. TEM studies of layers containing 6 at. % C also revealed crystalline material with the diamond-cubic structure although stacking faults and twins were also visible. No evidence of SiC precipitation was observed. Diffraction revealed a subtle decrease in the lattice constant of the unit cell with increasing C content which is consistent with the smaller size of C substituting for the larger size of Ge in the lattice. The remarkably low deposition temperature undoubtedly favors incorporation of C in the lattice as indicated by complementary IR and Raman analysis of the films. The Raman spectrum (Figure 5a) displays a weak absorption at 620 cm^{-1} corresponding to the localized mode of substitutional C in Si. Strong absorption peaks at 800 cm^{-1} that would correspond to SiC are not observed. Further characterization by SIMS reveals that the films are homogeneous and free of impurities (Figure 5b).

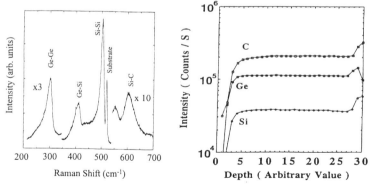

Figure 5a. Raman spectrum, representative of the entire sample, of $Si_{0.61}Ge_{0.32}C_{0.07}$. Absorption at 620 cm^{-1} corresponds to vibrational C modes in the Si lattice. Figure 5b. SIMS depth profiles of the same sample.

Our results confirm that incorporation of substantial C concentrations into substitutional sites is best achieved by low temperature deposition reactions involving C-H free precursors which provide crystalline layers with C varying from 4-6 at. %.

Precursors that contain C-H bonds require high deposition temperatures that favor SiC formation and often result in growth of amorphous materials. Since high C concentrations may be necessary to influence the band structure and produce materials that match the silicon lattice, we envisage that our synthetic approach will have substantial impact in the further useful development of the Si-Ge-C system.

Synthesis of Ge-C. A closely related but simpler binary material that has potential for Si-based bandgap engineering as well as lattice matching with Si is the $Ge_{1-x}C_x$ alloy system. This new material may contain compositions that have bandgaps larger than those of SiGe and these are expected to have important applications in high speed transistors and optoelectronics. In addition, the presence of C in the lattice should reduce the lattice mismatch between Ge and Si that prevents growth of pure Ge epitaxial layers on Si. This growth usually starts in a 2-D mode up to a certain thickness of typically a few monolayers. A transition to a 3-D island growth then occurs because the system is minimizing its inherent strain through island formation (Stranski-Krastanov growth). As and Sb surfactant-mediated growth of Ge suppresses island formation and appears to result in 3-D epitaxial growth (15). Apparently, the presence of surfactants on the growth front reduces surface diffusion in MBE experiments to eventually yield highly strained Ge epilayers that are considerably thicker than just a few monolayers.

$Ge_{1-x}C_x$ hybrids of diamond and Ge are much harder to prepare than $Si_{1-x-y}Ge_xC_y$ alloys because of the metastability of the Ge-C bond in the solid state with respect to Ge and C. As a result of this instability, the equilibrium solubility of C in Ge is negligible at all temperatures and pressures and a crystalline GeC analog of SiC has not yet been prepared. Routes to produce alloys as epitaxial thin films are of much interest because virtually nothing is known about $Ge_{1-x}C_x$ and current CVD technology has failed to produce crystalline films. Recently, Sb mediated growth of epitaxial $Ge_{1-x}C_x$ films with C concentrations of about 1 at. % by MBE was reported(16, 17). A more detailed investigation using similar techniques described deposition of Ge-C films on Si that contained small epitaxial domains embedded in an amorphous matrix (18). TEM analysis of the crystalline regions indicated a lattice parameter indistinguishable from that of pure Ge.

Our work has been directed towards the development of low temperature methods involving UHV-CVD and new precursors to grow crystalline Ge-C alloys with substantial C incorporation (5-10 at.%). This would allow us to study systematically the effects of varying carbon content on the lattice constant and the band structure. Since syntheses of heteroepitaxial $Si_{1-x-y}Ge_xC_y$ (y=0.01-0.06) alloys by CVD have utilized silylmethanes as precursors, analogous germylmethanes might be ideal precursors for deposition of $Ge_{1-x}C_x$ that contain substantial C concentrations. We have explored reactions of GeH_4 with $(GeH_3)_3CH$ to deposit monocrystalline and epitaxial Ge/C alloys with C concentrations of up to 6 at. %.

Interactions of $(GeH_3)_3CH$ with GeH_4 at 450°C on Si produced virtually perfect heteroepitaxial layers (300-1000Å thick) of 2-D Ge-C alloys with high C content. The C concentration in all samples was determined by RBS analysis utilizing the resonance reaction at 4.28 MeV. This technique enhances the C signal by 128-fold thus making it possible to quantify C contents as low as 0.5 at.%. We took special care to differentiate the bulk C signal from the surface C in order to obtain the most accurate measurement. Figure 6 shows typical resonance spectra for $Ge_{1-x}C_x$

modeled using the program RUMP to determine the C composition. We demonstrated by TEM, EELS, and RBS ion channeling that C primarily occupied substitutional sites in the diamond-like Ge lattice. For example, the electron micrograph in Figure 6b shows the heteroepitaxial growth of $Ge_{1-x}C_x$ (x=0.020) on Si. Electron diffraction indicated a lattice constant slightly smaller than that of Ge. The lattice constant of samples with composition $Ge_{0.95}C_{0.045}$ was 5.54 Å, about midway between those of Ge and Si.

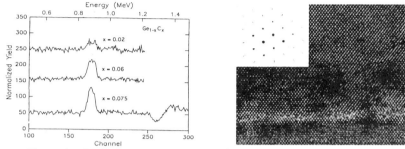

Figure 6a. (RBS) C resonance spectra of $Ge_{1-x}C_x$. Figure 6b. High-resolution electron micrograph showing cross section of epitaxial $Ge_{0.98}C_{0.02}$.

Further TEM observations of Ge-C with C concentrations from 1.5-6.0 at. % revealed that the layers were crystalline and heteroepitaxial, but with {111} stacking faults and microtwins originating at the Ge-C/Si interfaces (19-20). Channeled spectra of selected samples showed a substantial decrease of the C signal relative to the random signal, indicating that C occupied substitutional sites in the Ge lattice (Figure 7). EELS spectra confirmed the presence of C and suggested that it is sp^3 hybridized. Analysis of areas less than 20 Å revealed the K-shell edge of C featuring a peak attributable to σ* transitions which are characteristic of diamond-like C. Transitions corresponding to π* excitations which are due to graphitic C were not present in the spectrum.

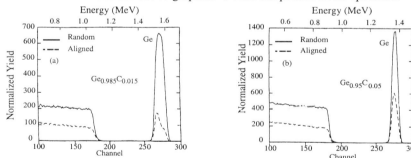

Figure 7. The 2 MeV random and aligned spectra for $Ge_{1-x}C_x$ (x= 0.015, 0.05). Dechanneling observed near interface suggests the presence of misfit dislocations. Channeling decreased with increasing C content.

During the deposition of the $Ge_{1-x}C_x$ films we also observed a substantial growth rate decrease as the C concentration increased (ranging from 5-6 Å per minute for 1.5 %C to 2 Å per minute for 5-6 % C, which provided 2-D layer-by-layer growth which was important for epitaxial growth of our materials. We investigated the film morphology vs. C content and growth rate by TEM and atomic force microscopy (AFM). We found that even C concentrations as low as 1.5-2 % have substantial

122

effect on the structure, morphology, and the growth behavior. For comparison, we performed control experiments involving growth of pure Ge on Si at conditions identical to those used for deposition of $Ge_{1-x}C_x$. As expected, the experiments resulted in thick, incommensurate layers of Ge via a 3-D island-like growth mechanism. Typical rates ranged from 45 to 55 Å per minute whereas those for $Ge_{0.95}C_{0.05}$ were substantially lower at approximately 2 Å per minute. AFM examination of the surface morphologies indicated that the pure Ge films were extremely rough and the top of the layer was dominated by large island-like domains. The AFM image of a sample containing 1.5 % C revealed that the film surface was considerably smoother and that the island formation was suppressed considerably. The average roughness (RMS) was reduced to just a few monolayers. The surface roughness improved even further with increased carbon content. Samples containing 3 % C displayed virtually atomically flat surfaces as shown by AFM and cross-sectional TEM examinations. AFM images of a Ge film on Si and $Ge_{1-x}C_x$/Si samples for x=0.015 and x=0.03 displaying the dramatic change in surface morphology are shown in Figure 8a. A representative TEM micrograph of $Ge_{0.95}C_{0.05}$ demonstrating a smooth and continuous surface is shown in Figure 8b.

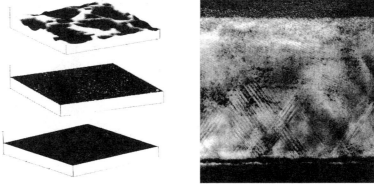

Figure 8a. AFM images showing: Ge film on Si and $Ge_{1-x}C_x$ for x= 0.015 and x= 0.030. Figure 8b. TEM micrograph showing smooth surface of $Ge_{0.95}C_{0.05}$.

Group IV semiconductor compounds. Ordered structures composed from Si, Ge, C constituents offer the possibility of minimizing any strain energies by ordering the constituents in the same way that 50 at % Si and 50 at % C are ordered in the stable zincblende SiC compound. A recent theoretical investigation has examined a variety of hypothetical compounds with the diamond, zincblende, copper-gold, copper platinum, chalcopyrite, and luzonite crystal structures (21). With the exception of SiC, all systems considered were found to be metastable, although several show promise for synthesis. The Si-Ge-C compounds that have C on one common sublattice and Si and Ge on another appear to be the most stable and are nearly lattice-matched with SiC. We are involved in synthesis of strain-stabilized diamond-like and zinblende ordered structures which incorporate the carbon on a common sublattice.

Synthesis of GeSi3C4. We have recently demonstrated the synthesis of a new Si-Ge-C material with a composition $GeSi_3C_4$ (11) and a structure related to ZnS sphalerite rather than diamond-cubic. The diffraction data and phase composition suggest a face-centered-cubic ordered structure as shown in Figure 9.

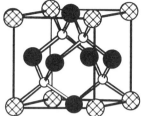

Figure 9. Zincblende based crystal structure of Si_3GeC_4. Shaded spheres at corners and black faces represent Ge and Si atoms respectively. Small spheres at T_d sites represent carbon atoms.

The material was obtained by thermal decomposition (650-750°C) of $(SiMe_3)_4Ge$, which is a volatile air-stable compound, via elimination of CH_4, H_2, and $HSiMe_3$. Growth rates of 20 nm per minute were observed and the conversion of reactant to product was nearly complete. Compositional, structural and vibrational analyses revealed that this was a single phase containing Ge-C bonds. The compositions of thin films deposited on Si and SiO_2 were determined by RBS to be $Si_{0.37}Ge_{0.13}C_{0.50}$. Diffraction studies of samples on Si revealed polycrystalline material with the SiC cubic structure but possessing a slightly larger lattice constant. IR studies revealed Si-C and Ge-C stretches at 800 cm^{-1} and 720 cm^{-1}, respectively, which are absorptions indicative of the zincblende structure. Overall, this precursor provides an extremely efficient method for depositing cubic Si-Ge-C. The material developed is expected to be of value in high-temperature electronics and possibly also serve as a buffer layer for the deposition of epitaxial SiC.

Synthesis of $(Si_2Ge)C_x$. We have observed a novel ordered atomic arrangement during formation of diamond-like Si-Ge-C layers with composition $(Si_2Ge)_{32}C_4$. The microstructure of this material is dominated by two light rows of lattice fringes and one dark row along the [111] direction as visible in figure 10.

Figure 10. TEM micrograph and diffraction pattern inset of $(Si_2Ge)_{32}C_4$.

Since these features dominate the structure, especially above the interface we postulate that it may consist of a superlattice in which two planes of Si follow a plane of Ge. We expect that C is randomly distributed in the Si planes because the

C(SiH₃)₄ precursor contains C tetrahedrally surrounded by Si atoms. The IR spectra of the material also show random substitutional C having four silicons as nearest neighbors. The diffraction pattern consists of the primary diamond cubic spots and it also shows two additional spots between each primary spot corresponding to superlattice reflections from the tripling of the [111] periodicity. The incorporation of C in this material appears to facilitate formation of this novel phase.

Synthesis of Si₄C. We have prepared a new group IV binary compound of composition Si₄C (22). This phase has been predicted to consist of Si₄C tetrahedra linked together in a diamond-like network in which the C atoms occupy a common sublattice (23). In theory, this ordered structure is energetically preferred over a random alloy because it allows the stretched Si-C bond to exist with minimum strain. The Si₄C phase was deposited by the decomposition of the precursor C(SiH₃)₄ in the presence of H₂ carrier gas at temperatures ranging from 550-700 °C. RBS analysis indicated that the elemental content was Si₈₀C₂₀ which is the same composition as the original Si:C ratio in the precursor. This result implies that the composition and potentially the tetrahedral structure of the CSi₄ core were retained in the solid state. As-deposited material did not contain any hydrogen as determined by RBS and confirmed by vibrational studies. Figure 11 shows the RBS spectra obtained from a sample deposited at 650 °C. From simulation of the spectra, the film thickness was estimated to be 45 nm, which corresponded to a growth rate of 0.24 nm/min.

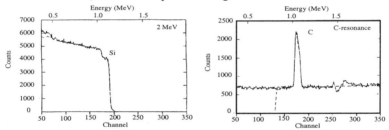

Figure 11. (a) 2 MeV RBS spectrum of film with composition $Si_{0.80}C_{0.20}$. (b) 4.3 MeV He $^{2+}$ carbon resonance spectrum of $Si_{0.80}C_{0.20}$.

TEM observations, (Figure 12a) revealed that the as-deposited film was completely amorphous. Samples annealed at T<750 °C showed partial crystallization at the substrate alloy interface whereas annealing at intermediate temperatures 800°C <T< 850°C led to complete crystallization across the entire layer thickness and some limited interdiffusion at the substrate-layer interface Figure 12b. The material was single phase with the diamond-cubic structure (no evidence for SiC precipitation was found) although extensive lattice defects were observed. Diffraction indicated a lattice constant ranging locally from ~0.540nm to ~0.530nm, which is slightly lower than that of Si (0.543nm) but significantly higher than the value of 0.506 nm calculated from Vegards' law by assuming linear interpolation of the unit-cell parameters of diamond -C and Si. RBS of the annealed samples confirmed that the composition had remained as SiC. SIMS depth profile experiments showed that the Si and C were still homogeneously distributed throughout the sample. Complementary EELS studies confirmed the presence of C in the layer and also showed that the carbon was sp³ hybridized and part of the diamond-cubic structure.

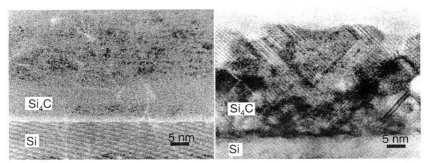

Figure 12a Electron micrograph of amorphous $Si_{0.80}C_{0.20}$ as-deposited at 550 °C. Figure 12b Micrograph of $Si_{0.80}C_{0.20}$ annealed at 850 °C. Complete crystallization via solid phase epitaxy has occurred across the entire layer.

We found that the Si_4C phase only grew as a thin crystalline epitaxial layer to a maximum thickness of about 40nm, above which the system became polycrystalline. Lattice parameters as low as 0.525nm were measured in small grains, suggesting that the larger lattice parameter measured in the epitaxial layers may have been due to substrate-induced strain. Semiconductor structures that are not found in the bulk phase diagram have been previously shown to be stabilized by substrate-induced strain. (24) Larger-than-normal Si-C bonds and possible strain stabilization of the Si_4C phase are consistent with the proposed theoretical description of the structure. At T>950°C nucleation of crystalline particles occurred which were identified as $Si_{1-x}C_x$ and β-SiC precipitates, indicating the instability of Si_4C with respect to the thermodynamically stable SiC at higher temperatures.

The FTIR spectra from the as-deposited sample showed only the C-Si stretching mode as a broad absorption peak centered at 730 cm^{-1}, which probably indicated a lack of long-range order. Comparison of the stretching frequency with that of cubic β-SiC at 800 cm^{-1} and the local vibrational mode in substitutional $Si_{1-x}C_x$ alloys at 610 cm^{-1} indicates that the Si-C bond is symptomatic of a weakened bond presumably due to steric strain. The molecular structure of the precursor determined by gas-phase electron diffraction (14) revealed that the SiC bond length was 1.88Å, nearly identical to that of SiC (1.89Å). Thus, the transition from the vapor phase to the solid state has caused substantial elongation and weakening of the Si-C bond, presumably to accommodate the strain arising from the sterically crowded Si environment. According to the proposed model structure (23), the C atoms in Si_4C are arranged as third-nearest neighbors, thereby accommodating the stretched C-Si bond which is 7% longer than the Si-C bond in β-SiC. We attribute the larger than expected lattice parameters as due partly to steric repulsions in the lattice causing substantial elongation of the SiC bonds.

Precursor Synthesis

Synthesis of $(BrCl_2Ge)_4C$ and $(Br_3Ge)_4C$. The complexes, GeX_2·dioxane (26) (X = Cl, Br), undergo complete insertion into the C-Br bonds of CBr_4 to give the compounds, $(BrCl_2Ge)_4C$ **1** and $(Br_3Ge)_4C$ **2**, in 79 % and 94 % yields, respectively

(Eqn. 1) (27). This is the first example of a germylene insertion reaction leading to complete substitution at a single carbon center. Compounds **1** and **2** are crystalline, air-stable, and high-melting point solids that behave similarly to other highly symmetric compounds containing a tetrahedral core of group IV-A elements (28). Spectroscopic characterizations as well as elemental analysis are consistent with the molecular formulas of **1** and **2**.

$$4 \; GeX_2 \cdot dioxane \; + \; CBr_4 \quad \xrightarrow[- \; dioxane]{toluene} \quad \text{(structure)} \quad (1)$$

(X = Cl, Br)

In order to determine the bonding parameters of the Ge-C interactions in these molecules, an X-ray crystallographic analysis was performed for **2**. However, the high symmetry of the molecule prevents it from ordering the four Ge positions within the crystal lattice. This internal disorder leads to a model in which the four Ge atoms are distributed amongst twenty partially occupied sites in a dodecahedral arrangement. The central location of the C atom allows it to refine with full occupancy. In addition, the tight packing of the Br atoms on the outer surface of the molecule fixes their positions capping each face of the dodecahedron. The $C_1Br_{12}Ge_{4.06}$ stoichiometry obtained by permitting the occupancy parameters of the Ge atoms to refine is extremely close to the true stoichiometry.

Although the X-ray data cannot provide the true structure of **2** (i.e. a central C tetrahedrally bound to four $GeBr_3$ moieties), the resulting Ge-C distances have a narrow range (2.006 - 2.051 Å) and should approximate the true Ge-C bond distances. These values are significantly longer than the normal Ge-C bond lengths which are close to 1.94 Å in carbogermanes. The unusual bond lengths found in **2** suggest that the molecules (and crystals) with a central C bonded to four Ge might be sterically crowded. Accordingly, we undertook a precise determination of the molecular structure of **2** by gas phase electron diffraction (Figure 13a) (29).

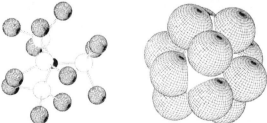

Figure 13a   Structure of $C(GeBr_3)_4$ with T symmetry.   The small dark sphere is C, and the colorless and shaded spheres represent Ge and Br atoms respectively.   Figure 13b   Packing model of the Br atoms in $C(GeBr_3)_4$.   Notice that each Br atom has five Br neighbors in an icosahedral arrangement.

The structure refinement was based on a model of T symmetry characterized by four parameters: the Ge-C and Ge-Br bond distances, the C-Ge-Br angle and the torsion dihedral angle τ(Ge-C-Ge-Br). There are two striking features of the structure to which we call attention. First the remarkably long Ge-C bond length of the disordered crystals found by X-ray diffraction has been confirmed by the experimental molecular structure (Ge-C=2.048 (8) Å). It is fully 0.10 Å longer than a typical Ge-C bond and it is symptomatic of a weakened bond presumably due to steric strain. The Ge-C bond in carbogermanes is about 1.94 Å as illustrated in the following examples: d (Ge-C) 1.945 Å in $Ge(CH_3)_4$, 1.947 Å in $GeH(CH_3)_3$, 1.950 Å in $GeH_2(CH_3)_2$ and 1.945 Å in GeH_3CH_3. Second, as it should be apparent from (Figure 13b), the arrangement of the Br atoms are close to icosahedral. An icosahedral arrangement of the 12 Br atoms maximizes the Br-Br distances and minimizes steric repulsions. Presumably this is the reason for this geometry. Comparison with bond distances in GeH_2Br_2 [2.277(3) Å] and $GeBr_4$ [(2.272(3) Å] indicates that the Ge-Br bond distances in $C(GeBr_3)_4$ [2.283(3) Å] are normal. We note that the idea of steric repulsion elongating bonds and reducing the stability of tetrahedral molecules is not new. In 1981 Tolman et al. remarked (30),"one suspects that $C(SiH_3)_4$ and $C(GeH_3)_4$ might not be stable for simple steric reasons" In fact they are stable and their utility as precursors in producing novel diamond structured phases has been demonstrated.

Synthesis of (H3Ge)4C and (H3Ge)3CH. The reduction of **1** and **2** with LiAlH4 has produced $(H_3Ge)_4C$ **3** in 20 % and $(H_3Ge)_3CH$ **4** in 30 % yield (Eqn. 2). Compounds **3** and **4** are isolated as low-volatility liquids and characterized by IR, NMR, mass spectrometry and gas phase electron diffraction studies.

A more convenient synthesis of $(GeH_3)_3CH$ **4** involves reduction of a novel methane $(GeBr_3)_3CH$. We synthesized $(GeBr_3)_3CH$ **5** in 85 % yield by complete insertion reactions of $GeBr_2$·dioxane into the C-Br bonds of $HCBr_3$ as shown below.

$$HCBr_3 + 3\ GeBr_2\cdot dioxane \longrightarrow (Br_3Ge)_3CH + 3\ dioxane \qquad (3)$$

Compound **5** is a colorless, air-sensitive solid, highly soluble in organic solvents that melts without decomposition at 150°C. Its identity has been established by FTIR, NMR, GC-MS, and elemental analysis. Reduction **5** with LiAlH4 (Eq.4) under conditions similar to those used in the synthesis of $(SiH_3)_4C$ [15] yield $(GeH_3)_3CH$ **4** in 20% yields.

$$4\ (Br_3Ge)_3CH + 9\ LiAlH_4 \longrightarrow 4\ (GeH_3)_3CH + 9\ LiBr + 9\ AlBr_3 \qquad (4)$$

Molecular structure of (H3Ge)3CH and (H3Ge)4C as determined by electron reduction. The gas phase molecular structure of $(H_3Ge)_3CH$ was obtained in order

128

to determine the Ge-C interactions in a less-crowded molecule (Figure 14a). The structure refinement was based on a model of C_3 symmetry, with local C_{3v} symmetry and gave a Ge-C bond length of 1.96 Å. The bond distance is slightly longer than normal Ge-C bond distances (1.945 Å in carbogermanes). Some elongation possibly due to steric repulsions is suggested which is also demonstrated by a very small widening of the C-Ge-C angle (111.0) from the tetrahedral value. Structure optimization of $(H_3Ge)_4C$ was carried out with a model of T symmetry (Figure 14b). The best value obtained for the dihedral GeCGeH angle was 162° indicating that the germyl groups have been rotated 18.2° away from the staggered T_d orientation. The Ge-C bond distance, 1.97 Å, is slightly longer by only 0.02 Å than in typical carbogermanes and indicates that the compound is relatively free of strain. This is in strong contrast with the Ge-C bond length obtained from the gas phase structure of the $C(GeBr_3)_4$ (2.049 Å).

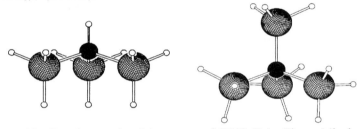

Figure 14a Gas-phase molecular structure of $CH(GeH_3)_3$. Figure 14b shows the gas phase molecular structure of $C(GeH_3)_4$

Conclusion

We have used germylene chemistry to construct molecules containing a central carbon tetrahedrally encapsulated within a germanium environment. The compounds are stable despite the large amount of steric bulk, suggesting that substitutional carbon in a Ge cubic lattice may be able to exist without creating strain defects. Furthermore, their reduction with $LiAlH_4$ leads to the formation of $(H_3Ge)_3CH$ and $(H_3Ge)_4C$ crucial precursors for UHV-CVD applications and bandgap engineering studies.

We have achieved C incorporation into Si-Ge by reactions of $C(SiH_3)_4$ or $C(GeH_3)_4$ with GeH_4 and SiH_4 to yield metastable diamond-like $Si_{1-x-y}Ge_xC_y$ (y=0.04-0.07). Synthesis of ordered phases such as Si_4C, $SiGe_3C_4$, and $(Si_2Ge)C_x$ was also demonstrated. These materials are energetically preferred over the alloys and offer the prospect of having bandgaps wider than that of Si and in some cases the bandgaps are expected to become direct. Low temperature growth of crystalline $Ge_{1-x}C_x$ is now possible by using the unique combination of UHV-CVD and chemical precursors. The island-like 3-D nucleation observed during growth of Ge films has been suppressed by the incorporation of carbon.

Acknowledgment

This work was supported by NSF, Grant No. DMR 9458047

References

1. Patton, G. L.; Harame, D. L.; Strock, J. M.; Meyerson, B. S.; Scilla, G. S. *IEEE Electron Device Letters* **1989**, *10*, 534.
2. Patton, G. L.; Comfort, J. H.; Meyerson, B. S.; Crabbe, E. F.; Scilla, G. J.; Strock, J. M. C.; Sun, J. Y. C.; Harame, D. L.; Burghartz, J. N. *IEEE Electron Device Letters* **1990**, *11*, 171.
3. Meyerson, B. S. *IBM J. Res. Develop.* **1990**, *34*, 806.
4. Hoffman, L.; Bach, J. C.; Nielsen, B. B.; Leary, P.; Jones, R.; Oberg, S. *Phys. Rev. B* **1977**, 11167
5. Soref, R. A. *J. Appl. Phys.* **1991**, *70*, 2470.
6. Demkov, A.A.; *Phys. Rev. B* **1993**,48, 2207
7. Rim, K.; Takagi, S.; Welser, J. J.; Hoyt, J.L.; Gibbons, J. F. *Mat. Res. Soc. Symp. Proc* **1995**, 379, 327
8. Chang, C.L.; Amour A. St.; Strum, J. C *Appl. Phys. Lett.*,**1997**, 70,1557
9. Powell, A. R.; Eberl, K.; Ek, B. A.; Iyer, S. S. *J. Cryst. Growth* **1993**, *127*, 425.
10. Strane, J. W.; Stein, H. J.; Lee, S. R.; Doyle, B. L.; Picraux, S. T.; Mayer, J. W. *Appl. Phys. Lett.* **1993**, *63*, 2786.
11. Kouvetakis, J.; Todd, M.; Chandrasekhar, D.; Smith, D. J. *Appl. Phys. Lett.* **1994**, *65*, 2960.
12. Todd, M.; Matsunaga, P.; Kouvetakis, J.; Smith, D. J. *Appl. Phys. Lett.* **1995**, *67*, 1247.
13. Iyer, S.S.; Eberl, K.;Goursky, M. S.; LeGoues, F.K.; J. C. Tsang, J. C.: *Appl. Phys. Lett.* 1992, **60**(3), 356
14. Hager, R.;Steigelman, O.; Muller, G.; Schmidbaur, H.; H. Robertson, H. H.; Rankin, D. W. *Angew. Chem. Int. Ed.* **1990, 29**(2), 201;15.
15. Osten, H. J.; Bugiel, E.; Zaumseil, P. *J. Cryst. Growth* **1994**, *142*, 322.
16. Osten, H. J.; Klatt, J. *Appl. Phys. Lett.* **1994**, *65*, 630.
17. Kolodzey, J.; O'Neal, P. A.; Zhang, S.; Orner, B. A.; Roe, K.; Unruh, K. M.; Swann, C. P.; Waite, M. N.; Shah, S. I. *Appl. Phys. Lett.* **1995**, *66*, 1865.
18. Krishnamurthy, M.; Drucker, J. S.; Challa, A. *J. Appl. Phys.* **1995**, *78*, 7070.
19. Todd, M.; Kouvetakis, J.; Smith, D. J.; *Appl. Phys. Lett.* **1996**, 395, 79
20. Todd, M.; Kouvetakis, J.; McMurran, J.; Smith, D. J.*Chem. Mater.* **1996**, 8, 2491
21. Berding, M. A.; Sher, A.; Schilfgaarde, van, M. *Phys. Rev. B*, in press
22. Kouvetakis, J.; Smith, D. J.; Chandrasekhar, D.; *Appl. Phys. Lett.* **1998**, 72, 930
23. Newman, R. C.; Willis, J. B. J. *Phys, Chem, Solids* **1965**,26, 373
24. (a) Martins, J. L.; Zunger, A. *Phys. Rev. Lett.* **1986**, 56, 1400. (b) Flynn, C. P. *Phys. Rev Lett.* **1986**, 57, 599
25. Rucker, H.; Methfessel, M; Bugiel, E.; Osten, A. J. *Phys. Rev. Lett.* **1994**, 72, 3578
26. Hermann, W.; Denk, M., Eur. Patent 568074.
27. Matsunaga, P. T.; Kouvetakis, J.; Groy, T. L. *Inorg. Chem.* **1995**, *34*, 5103.
28. Kaczmarczyk, A.; Millard, M.; Nuss, J. W.; Urry, G. J. *Inorg. Nucl. Chem.* **1964**, 26, 421
29. Kouvetakis, J; O'Keeffe, M. *Inorg. Chem.* in Press
30. Tollman, J. J.; Frost, A. A.; Topiol, S.; Jacobson, S.; Ratner, M. A. *Theor. Chim. Acta.* **1981**, 58, 285

Chapter 10

The Chemical Vapor Deposition of Metal Boride Thin Films from Polyhedral Cluster Species

John A. Glass, Jr., Shreyas S. Kher, Yexin Tan, and James T. Spencer[1]

Department of Chemistry and the W. M. Keck Center for Molecular Electronics, Syracuse University, Syracuse, NY 13244–4100

Metal boride thin film fabrication through Chemical Vapor Deposition (CVD) techniques is an area of current technological and scientific interest. These materials have attracted significant attention due to their breadth of unique physical, chemical and structural properties. CVD is clearly one of the best methods for the formation of these materials since it circumvents many of the problems associated with other technologies. We have used several boron CVD precursors, especially the boranes, to prepare many pure metal and metal boride thin film materials through an apparently general deposition reaction process. In this paper, our work on the formation of polycrystalline transition metal and lanthanide metal boride thin films from borane precursors is briefly presented. These deposited materials have been extensively characterized by techniques including scanning electron microscopy (SEM), Auger electron spectroscopy (AES), X-ray diffraction (XRD), X-ray emission spectroscopy (XES), low-energy electron diffraction (LEED), transmission electron spectroscopy (TEM) and solid state nuclear magnetic resonance spectroscopy (NMR). This CVD chemistry appears to constitute a highly efficient method for the formation of polycrystalline transition metal and lanthanide metal boride thin films.

The fundamentally important chemical processes and reactions in the CVD formation of metal boride films from boranes has not, however, been previously well investigated. In our recent work which will be briefly summarized here, we have explored several of the important details of the CVD of metal borides from gas phase boranes and metal halide precursors.

[1]Corresponding author.

I. Introduction

The study of polyhedra, many-faced solids, has long intrigued and fascinated scientists and philosophers. Plato and Archimedes both devoted a great deal of study to these solids. Plato first described a series of five "pure" polyhedral bodies from which Archimedes later elegantly derived thirteen semi-regular polyhedra. Since these early works, polyhedral structures have continued to be the focus of a great deal of attention from philosophers, mathematicians and physical scientists. The field of solid state boron chemistry, however, probably most closely ties together the abstract study of pure polyhedra with the physical and chemical world. In particular, boron containing solids may be thought of as the bridge between discrete small molecule behavior, with more localized bonding, and that of extended solid arrays, with extensively delocalized electronic structures. Solid state metal boride materials, for example, frequently display structural and bonding features from both of these chemical behavioral schemes.

Among solid state materials, the metal borides are remarkable due to a combination of unique compositional and structural features, physical properties and potential applications to a wide variety of technological problems. The element boron not only combines with most metals but frequently does so to form a series of discrete binary phases with up to eight different metal-to-boron ratios for any given metal.[1] Metal borides are known with solid state structures which range from essentially isolated boron atoms to boron-boron bonded chains, two-dimensional continuous networks and complex three-dimensional frameworks which extend throughout the entire crystal.[2] The structure of neodymium hexaboride (NdB_6), shown in Figure 1, is an example of this latter type of complex three-dimensional structure in which B_6 octahedra are arranged in a body centered cubic lattice with the octahedra linked to the apices of other octahedra in all six directions, giving a rigid but yet relatively open structure. The strong multicenter, covalent bonding within these boron polyhedra is believed to impart the observed high stability,

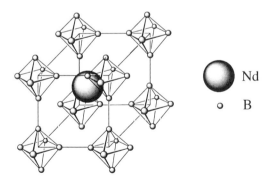

Figure 1. Structure of Neodymium hexaboride.

hardness and high melting points to the boride materials.[3] While it is not possible to entirely account for the boride structures in simple localized bonding terms, it is generally believed that the metal center donates electrons to the boron units in the boron rich compounds. In the case of materials such as NdB_6, the *closo*-boron octahedra require 14 valence electrons (2n+2), of which 12 are provided by the boron atoms.[4] If the neodymium then provides two electrons to the cage, one "free" valence electron should remain per metal center, making the material an excellent conductor. This analysis is consistent with Hall effect, solid state ^{11}B NMR and conductivity measurements on these materials.[5] Also supportive of this analysis is the fact that MB_6 materials which contain only divalent metal centers, such as the alkaline earth hexaborides (i.e., CaB_6), are insulators rather than conductors, since no "free" electrons remain on the metal centers for conduction.[2] Thus, the bonding description of the complex metal borides, such as the trivalent hexaborides, may be thought to contain both delocalized covalent (within the polyhedra) and ionic bonding modes (between the polyhedra and the metal). The best electronic description of these materials, however, comes from a more complex molecular orbital treatment such as recently presented by Hoffmann for Ta_3B_4.[6] Several metal boride epitaxial thin films, such as HfB_2, have also recently been carefully studied by XPS and related techniques in efforts to probe experimentally the electronic structures of these materials more fully.[7] The unique structural and electronic diversity of the metal borides thus continues to inspire both theoretical and synthetic investigations of these materials. Indeed, several reviews of these solid state materials have appeared in which the literature has been presented in detail.[1]

The metal borides are typically very refractory materials, possessing very high melting points, exceptional hardnesses and high thermal electric conductivities. For example, the diborides of Zr, Hf, Nd, and Ta all have melting points of well over 3000° C, which exceeds those of the pure parent metals.[8] One additional important characteristic property of metal borides is that they possess electrical conductivities of a metallic order and several borides, such as LaB_6 and TiB_2, have electrical resistances very much lower than the corresponding pure metals (i.e., over five times lower for TiB_2). The metal borides also exhibit enormous thermal stability and are not attacked by dilute acids, bases or even concentrated mineral acids. Because of these properties, metal borides have found critical uses in a variety of applications, ranging from "low technology" hard cutting surface coatings to advanced optoelectronic systems. They are also of interest for potential application in patterned depositions on semiconductor substrates for use in high electron mobility transistors (HEMT), pseudomorphic and heterostructural devices, heterojunction bipolar transistors (HBT), and ultra-high speed microelectronic devices. The lanthanide metal borides have recently become the center interest due not only to their refractory, magnetic and electrical properties but also to their potential use as excellent thermionic materials. Lanthanum hexaboride, for example, has the highest electronic emissivity of any known material and its performance is unaffected by the presence of either nitrogen or oxygen.[9] Lanthanum hexaboride thin film cathodes have recently been successfully used to replace nickel cathodes in display panels. Finally, the metal boride materials, due to their very high thermal and high energy (10^4 - 10^6 eV) neutron capture cross sections, have been employed as neutron shields and in related "nuclear-hardened" applications.

The synthesis of solid state metal boride materials has employed a variety of preparative strategies.[2] Most of these methods, however, require very high temperatures (above 1000° C) and employ the use of low volatility precursors, such as metal oxides and boron or boron carbide.[10] None of the traditional methods for preparing metal borides, however, may be in any sense termed general. Because of the nature of the synthetic techniques and the refractory properties of the metal borides themselves, pure metal boride materials have been both difficult to prepare and analyze. In addition, the preparation of metal borides has focused almost entirely on the formation of bulk materials, rather than on the technologically important and scientifically interesting thin film materials.

The production of high quality thin films requires suitable materials chemistry and deposition pathways for the preparation of pure epitaxial structures. Chemical vapor deposition (CVD) methods have recently been shown to be among the most effective methods for the deposition of pure thin films.[11,12] CVD methods provide numerous advantages over traditional deposition methods including; (1) clean and controllable stoichiometric deposition processes in which film properties are primarily dependent upon the choice of precursors and easily controlled deposition conditions, (2) superior thin-film uniformity, (3) potential for pattern deposition with sharp boundary features through "real-time" processes, (4) lower temperature depositions, (5) significant migration reduction at film-substrate interfaces, (6) experimental ease of deposition from starting precursor materials, and (7) the facility for larger scale production processes. Most non-CVD processes are carried out at elevated temperatures, which frequently causes severe interlayer diffusion and results in vague interlayer junctures. In recognition of these advantages, considerable effort has been directed toward employing CVD techniques for forming semiconductor refractory thin-films.[11] These advantages make CVD chemistry the methodology of choice for the controlled formation of high quality metal boride materials.

The chemical vapor deposition of thin films of metal borides has previously presented significant challenges. The CVD of transition metal boride films from single-source metallaborane CVD precursors has recently been reported.[13,14] Complexes such as $[B_2H_6Fe_2(CO)_6]$, $[B_2H_6Fe_2(CO)_6]_2$, $[HFe_3(CO)_9BH_4]$, and $[HFe_3(CO)_{10}BH_2]$ have been used for the formation of iron boride thin films.[13] While not a CVD process, the synthesis of bulk gadolinium boride phases, such as GdB_4 and GdB_6, from a single molecular precursor, $Gd_2(B_{10}H_{10})_3$, at 1000-1200° C has also been recently reported.[15] In this report, powders containing both the gadolinium borides and amorphous boron were obtained as products from the thermolysis of the molecular precursor $Gd_2(B_{10}H_{10})_3$.[15] While the single source feature of this method may seem attractive, the deposition of these films, however, has typically lacked sufficient compositional control and the deposited materials were either amorphous or crystallized only after prolonged annealing. Most importantly, however, is the fact that essentially all metallaborane complexes are comparatively difficult and time consuming to prepare in pure form and in sufficient quantities for CVD applications.[16]

Transition metal borohydride complexes, such as $Ti(BH_4)(dme)$ and $Zr(BH_4)_4$,[17] have also been used as precursors in the CVD preparation of several metal boride thin films.[14,18] It appears that when the metal coordination sphere is completed solely by borohydride ligands, metal boride films result.[14,18-20] When hydridometalborohydride complexes are used instead, such as $AlH_2(BH_4)_3 \cdot 2N(CH_3)_3$, we found that extremely clean depositions of pure metal result.[19,20] The application of these precursors, however, is severely limited by both the extreme instability/reactivity and the synthetic difficulties encountered in the preparation of the metal borohydride precursor complexes. For example, lanthanaborohydride complexes are rare, with the neodymium and praseodymium borohydride complexes thus far entirely unknown. Most of the transition metal and lanthanide borohydride complexes which are known are thermally unstable, insoluble, intractable, nonvolatile solids, rendering them inappropriate for CVD methods.[17,21] Thus, the metal borohydride precursors are of only very limited potential for the formation of most metal boride thin films.

II. Experimental

Physical Measurements. Scanning electron micrographs (SEM) were obtained on an ETEC autoscan instrument in the N. C. Brown Center for Ultrastructure Studies of the S.U.N.Y. College of Environmental Science and Forestry, Syracuse, New York. Photographs were recorded on either Kodak Ektapan 4162 or Polaroid P/N 55 film. X-ray Emission Spectra (XES) were obtained on a Kevex 7500 Microanalyst System. The X-ray diffraction patterns (XRD) were recorded on a Phillips APD 3520 powder diffractometer equipped with a PW 1729 X-ray generator and a PW 1710 diffractometer control system. Copper $K\alpha$ radiation and a graphite single crystal monochromator were employed in the measurements reported here. The mass spectra were obtained on a VG 9000 glow discharge mass spectrometer using a 1 Torr argon discharge at 1 kV.

Materials. All solvents used were of reagent grade or better which, after appropriate drying, were degassed by repeated freeze-evacuate-thaw cycles and finally stored *in vacuo* prior to use.[22] *Nido*-decaborane(14), $B_{10}H_{14}$, was purchased from the Callery Chemical Company and was purified by vacuum sublimation before use. *Nido*-pentaborane(9), B_5H_9, was taken directly from our laboratory stock. The anhydrous metal halides were all commercially available and was used as received.

Chemical Vapor Deposition (CVD) of Metal Boride Thin Films. Typical CVD depositions were performed using a medium-high vacuum quartz reactor tube apparatus (1×10^{-6} Torr ultimate vacuum) employing a tube of 10 mm (o.d.) in diameter with a length of 60 cm (Figure 2). The apparatus was equipped with a chromel-alumel thermocouple with the thermocouple junction placed close to the tube in the middle in the oven. The reactor tube was placed horizontally in a tube furnace and was heated using an external electrical resistance furnace.

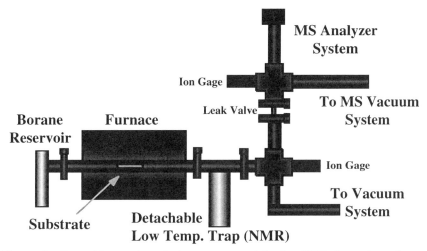

Figure 2. Deposition and analysis system used in the CVD formation of metal boride thin films from the vacuum copyrolysis of gas phase boron hydride clusters with metal chlorides.

The overall experimental operation of the reactor has been previously described.[23,24] In summary, however, in a typical experiment, 1.0 g (3.8 mmol) of an anhydrous metal chloride (such as $GdCl_3$, $FeCl_2$, $CoCl_3$, and $NiCl_2$) was placed in a quartz boat with the deposition substrates suspended over the top of the boat. The boat and substrates were then placed in the deposition system using inert atmosphere techniques.[22] The entire reactor system was evacuated to 4 x 10⁻⁶ Torr at room temperature for at least two hours prior to deposition. A boron precursor reservoir containing either freshly sublimed *nido*-decaborane(14), $B_{10}H_{14}$, or vacuum distilled *nido*-pentaborane(9), B_5H_9, was connected to the reactor. The borane reservoir flask was maintained at a constant temperature during the entire experiment by use of an external temperature bath jacketing the reservoir flask (typically from 22° C to 28° C for decaborane(14) and -78° C for pentaborane(9)). Control of the boron precursor flow into the reaction system was achieved through the use of narrow bore teflon vacuum stopcocks (0 to 4 mm) and by adjusting the temperature of the precursor flask by using an external constant temperature bath to modify its vapor pressure.[22] The reactor was then slowly heated under dynamic vacuum. After obtaining a stable temperature, the teflon valve to the borane reservoir flask was opened to allow a vapor of the borane to pass over the hot metal chloride while under dynamic vacuum conditions. The unreacted borane and other reaction by-products were trapped downstream in a liquid nitrogen-cooled trap. The deposition was continued for 1 to 3 h, during which time a film coated both the walls of the reactor and the deposition substrates held above the metal chloride boat. The stopcock to the borane flask was then closed and the reactor was allowed to cool slowly to room temperature. The reactor was filled with dry nitrogen and the film was removed from the system for further study.

III. Results and Discussion

Much of the research in CVD has until recently dealt primarily with the development of new deposition techniques and the purification of existent main group precursors. While these precursor compounds have proven adequate in some instances, the further development of advanced materials now relies upon the systematic design of precursors developed to exhibit enhanced chemical properties for deposition processes.

Our research in CVD chemistry involving metal boride materials, as summarized in this paper, has focused on the theoretical modeling of precursor structures and thermodynamics of the fragmentation pathways for source compounds, the design and chemical synthesis of new precursor compounds, the experimental study of decomposition pathways of sources compounds, and the CVD formation of high quality crystalline metal boride thin films.[11,12a, 19,20,22-24]

Recently, we have discovered that a variety of highly crystalline pure metal boride thin films may be prepared using CVD methods at relatively low temperatures through the vacuum copyrolysis of gas phase boron hydride clusters with metal chlorides. As a typical example, the formation of very high quality, polycrystalline thin films of neodymium, gadolinium and lanthanum hexaboride, LnB_6, was achieved through the pyrolysis of either nido-pentaborane(9) [B_5H_9] or nido-decaborane(14) [$B_{10}H_{14}$] with the corresponding lanthanide(III) chloride.[25] These films typically displayed deep blue colors, were very hard, and adhered very well to most deposition substrates. Depositions were carried out on a variety of substrates including quartz, copper, silicon, SiO_2, and ceramic materials. The lanthanide boride thin films were investigated by scanning electron microscopy (Figure 3), X-ray emission spectroscopy (Figure 4), X-ray diffraction, glow discharge mass spectrometry, and other techniques. X-ray diffraction and scanning electron microscopic data showed the formation of highly crystalline LnB_6 materials with a preferred orientation in the 111 direction. Spectroscopic data showed that the LnB_6 films prepared by these CVD techniques were very pure, highly crystalline and uniform in composition in the bulk material. Attempted depositions of gadolinium boride films on CaF_2 (111) resulted in the apparent formation of a ternary $(Ca/Gd)B_6$ phase in which the calcium is presumably substituted for gadolinium atoms in the cubic GdB_6 structure.[23]

In an effort to extend the use of borane-based source materials to the preparation of other thin films by CVD, we have also investigated the formation of transition metal boride films.[23,25] As a transition metal example, nickel boride was readily prepared with stoichiometric control from the pyrolytic reaction of $NiCl_2$ with several boron clusters.[23] The films were shown by AES to be compositionally uniform in the bulk sample as seen in the representative spectra presented in Figure 6. SEM data for the annealed boron-rich films, presented in Figure 5, showed the formation of perfect hexagonal crystals in a channeled columnar matrix. Electron

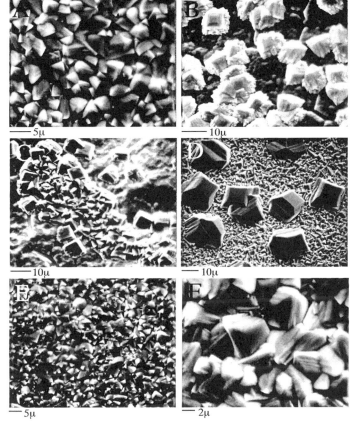

Figure 3. Scanning electron micrographs (SEM) of neodymium boride (NdB_6) films deposited from neodymium(III) chloride ($NdCl_3$) and *nido*-pentaborane(9) (B_5H_9) at 850° C on a quartz substrate. The bars below each photograph indicate scale.

diffraction data demonstrated that this hexagonal crystalline phase was a metastable orthorhombic Ni_7B_3 phase isolated in a Ni_3B matrix. The as-deposited nickel-rich films were found by XRD studies to be pure nickel with varying amounts of Ni_3B. Vacuum annealing of these nickel-rich films resulted in an observed decrease in the Ni_3B phase relative to the pure nickel phase in the XRD (Figure 7). We have also investigated the magnetic properties of these thin films by torque magnetometry and found them to be excellent magnetic media.

Table 1 gives several other examples of the formation of metal boride phases from metal chloride precursor compounds. Our survey of selected metal systems to date has found that, in every instance investigated, polycrystalline boride

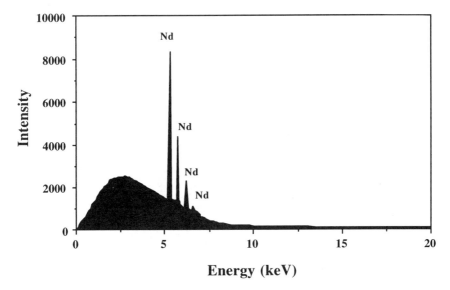

Figure 4.   X-ray emission spectrum (XES) of a typical NdB$_6$ film deposited from neodymium(III) chloride (NdCl$_3$) and nido-pentaborane(9) (B$_5$H$_9$) at 850° C on a quartz substrate.

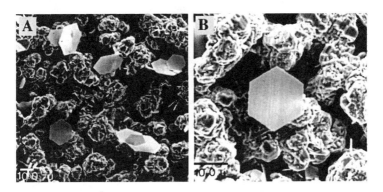

Figure 5. Scanning electron micrographs (SEM) of nickel boride thin films deposited at 502° C from NiCl$_2$ and *nido*-B$_5$H$_9$ on fused silica (quartz) and annealed at 830° C for 38 h post-deposition.[22] (b) shows and enlargement of one of the Ni$_7$B$_3$ hexagonal crystals from (c) in the Ni$_3$B matrix deposited from NiCl$_2$ and *nido*-B$_5$H$_9$.[23]

Table 1. Relationship between the metal precursor compound employed with deposited metal boride thin film material

Metal Precursor	Deposited Material
$TiCl_4$	TiB_2/Ti
$CrCl_3$	CrB_4
$MnCl_2$	MnB_4
$FeCl_2$	Fe_3B
$CoCl_2$	Co_2B/Co
$Cu(I)Cl$	Cu
$LnCl_3$	LnB_6
(Ln = lanthanide)	

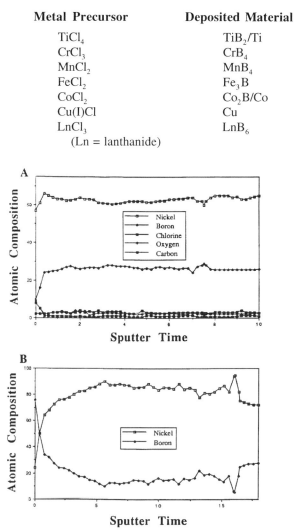

Figure 6. Representative Auger electron spectra for nickel boride thin films deposited using a *nido*-pentaborane(9), B_5H_9, borane source kept at -78° C during the deposition. The depth profiles were constructed from Auger electron spectra as the film was sputtered using Ar+ ion milling. Sputter times are given in minutes and the atomic compositions are given in percent. (a) Nickel boride film deposited at 500° C on pyrex. (b) Nickel boride film deposited at 497° C on fused silica (quartz).

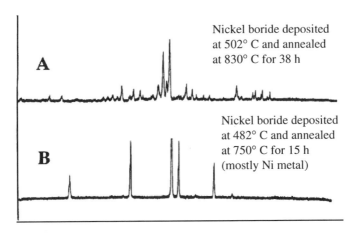

Figure 7. Representative X-ray diffraction spectra for nickel boride thin films deposited on fused silica (quartz) from NiCl$_2$ and *nido*-B$_5$H$_9$. The XRD spectra were recorded at room temperature from free standing films mounted on pyrex plates.

films were readily prepared. In all cases, the deposited materials were found to be conformal, polycrystalline and very hard.

In our work, we have employed two borane clusters as boron source compounds, either *nido*-pentaborane(9) [B$_5$H$_9$] or *nido*-decaborane(14) [B$_{10}$H$_{14}$]. Both of these precursors are *nido*-clusters which contain one open face.[4] These species have made particularly convenient source compounds since both are readily available, volatile and relatively easily handled. *Nido*-pentaborane(9) is, however, more easily controlled in a flow system because of its higher vapor pressure (v.p.$_{(25° C)}$ = 209 Torr)[26a] while *nido*-decaborane(14) is more easily handled overall since it is an air-stable material at room temperature (v.p.$_{(60° C)}$ = 1 Torr).[26b,c,27] Both of these clusters, however, have provided essentially identical metal boride depositions.

While we have demonstrated the highly efficient formation of polycrystalline metal boride thin films for a number of transition and lanthanide metals from borane precursors through CVD chemistry, the fundamentally important chemical processes and reactions involved in the depositions have, however, remained mostly a mystery. We have, therefore, begun investigations into the important chemical processes responsible for the deposition of metal boride materials. From previous work in our group, the nickel boride system was identified as one of the best experimental systems for detailed study.[23] Thus, we have focused our primary attention so far upon the elucidation of nickel boride CVD chemistry.

In studying the deposition pathways that are responsible for metal boride formation, an *in situ* probe is preferable, but *ex situ* experimental techniques are also currently proving particularly useful. Suitably modified experimental techniques such as UV-vis spectroscopy, mass spectrometry and gas chromatography are currently being extensively employed by many research groups in fundamental CVD investigations.[11] For gas phase processes, mass spectrometry is an excellent method for monitoring these reactions. In our work with nickel boride depositions, the vapor phase products were monitored primarily by mass spectrometry and ^{11}B NMR while the solid state products were typically analyzed by X-ray diffraction, SEM and TED.

In a typical deposition reaction, HCl, BCl_3, H_2BCl, $HBCl_2$, Cl_2 were detected as volatile species during the reactions of decaborane(14) with nickel (II) chloride. X-ray diffraction data for the films formed on the walls of the quartz reactor showed only Ni and Ni_3B phase with some very small other unidentified peaks.

The important chemical mechanistic steps which we have found from a variety of experiments for the CVD of nickel boride thin films are summarized in Figure 8. Briefly, the process begins with the pyrolytic deposition of boron hydride polymeric materials on the surface which then react with the nickel halide precursor to form the initially deposited boride material. Upon continued deposition, a "scavenger" reaction may occur in which some of the deposited boron may be removed from the metal boride material by eliminating BCl_3. While the overall process is reasonably complex and many details remain unknown, it appears that the major chemical reactions which contribute to the observed gas phase and solid state materials have been identified.

We are currently continuing with our exploration of the synthesis of new transition metal and rare earth borides by CVD and taking a more detailed look at the important chemical processes responsible for the observed metal boride deposition reactions.

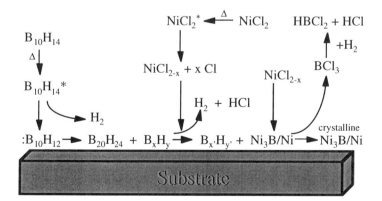

Figure 8.   Experimentally observed mechanistic steps in metal boride formation from borane clusters and metal halides.

142

IV. References

1. (a) Schwarzkopf, P.; Kieffer, R.; Leszynski, W.; Benesovsky, K. *Refractory Hard Metals, Borides, Carbides, Nitrides, and Silicates*; MacMillan Company: New York, 1953. (b) Aronsson, B.; Lundström, T.; Rundqvist, S. *Borides, Silicides and Phosphides*; Wiley: New York, 1965. (c) Pearson, W. B. *Crystal Chemistry and Physics of Metals and Alloys*; Wiley: New York, 1964. (d) Wells, A. F. *Structural Inorganic Chemistry*; Clarendon Press: Oxford, 1975. (e) Lipscomb, W. N. *J. Less-Common Met.* **1981**, *82*, 1. (f) Villars, P.; Carvert, L. D. *Pearson's Handbook of Crystallographic Data for Intermetallic Phases*; American Society for Metals: Metals Park, OH, 1985; Vol 2. (g) Samsonov, G. V.; Markovski, L. Ya.; Zhigach, A. F.; Valyashko, M. B. *Boron, Its Compounds and Alloys*; AN Ukr SSR Publishers: Kiev, 1960. (h) *Boron, Metallo-boron Compounds and Boranes*; Adams, R. M., Ed.; Interscience: New York, 1972. (i) *Boron and Refractory Borides*; Matkovich, V. I., Ed.; Springer-Verlag: New York, 1977. (j) Samsonov, G. V.; Goryachev, Yu. M.; Kovenskaya, B. A. *J. Less-Common Met.* **1976**, *47*, 147. (k) Kuz'ma, Yu. B. *Crystallography of Borides*; L'vov University Publishers: L'vov, USSR, 1983. (l) Hyde, B. G.; Andersson, S. *Inorganic Crystal Structures*; Wiley: New York, 1989.
2. Thompson, R. *Prog. Boron Chem.* **1970**, *2*, 173.
3. Johnson, R. W.; Daane, A. H. *J. Phys. Chem.* **1961**, *65*, 909.
4. (a) Mingos, D. M. P. In *Inorganometallic Chemistry*; Fehlner, T.P., Ed.; Plenum: New York, 1992; Chapter 4. (b) Wade, K. *Adv. Inorg. Chem. Radiochem. B*, *18*, 1.
5. (a) Lipscomb, W. N.; Britton, *J. Chem. Phys.* **1960**, *33*, 275. (b) Longuet-Higgins, H. C.; Roberts, M. De V. *Proc. Roy. Soc.* **1954**, *A224*, 336. (c)) Longuet-Higgins, H. C.; Roberts, M. De V. *Proc. Roy. Soc.* **1954**, *230*, 110. (d) Johnson, R. W.; Daane, A. H. *J. Phys. Chem.* **1963**, *38*, 425. (e) Silver, A. H.; Bray, P. J. *J. Chem. Phys.* **1960**, *32*, 288. (f) Lafferty, J. M. *J. Appl. Phys.* **1951**, *22*, 299.
6. Minyaev, R. M.; Hoffmann, R. *Chem. Mater.* **1991**, *3*, 547.
7. Belyansky, M.; Trenary, M. *Chem. Mater.* **1997**, *9*, 403.
8. Greenwood, N. N.; Earnshaw, A. *Chemistry of The Elements*; Pergamon Press: Oxford, 1985.
9. (a) J. H. Westbrook, *Proceedings of the International Symposium on High Temperature Technology*, New York, McGraw-Hill (1960). (b) A. F. Carter and G. P. Wood, NASA Memo-2-16-59L (1959).
10. (a) Paderno, Yu. B.; Fomenko, V. S.; Samsonov, G. V. *Ukrain. Khim. Zhur.* **1960**, *26*, 409. (b) Portnoi, K. I.; Timofeev, V. A.; Timofeeva, E. N. *Izvestiya Akademii Nauk SSSR, Neorganicheskie Materialy* **1965**, *1*, 1513.
11. Spencer, J. T. *Prog. Inorg. Chem.* **1994**, *41*, 145.
12. (a) Dowben, P. A.; Spencer, J. T.; Stauf, G. T. *Mat. Sci. Eng. B* **1989**, *B2*, 297. (b) Leys, M. R. *Chemtronics* **1987**, *2*, 155. (c) Davies, G. J.; Andrews, D. A. *Chemtronics* **1988**, *3*, 3. (d) Jones, A. C.; Roberts, J. S.; Wright, P. J.; Oliver, P. E.; Cockayne, B. *Chemtronics* **1988**, *3*, 152.
13. (a) Amini, M. M.; Fehlner, T. P.; Long, G. J; Politowski, M. *Chem. Mater.* **1990**, *2*, 432. (b) Thimmappa, B. H. S.; Fehlner, T. P.; Long, G. J.; Pringle, O. A. *Chem. Mater.* **1991**, *3*, 1148. (c) Jun, C.S.; Fehlner, T. P.; Long, G. J. *Chem. Mater.* **1992**, *4*, 440.

14. Jensen, J. A.; Gozum, J. E.; Pollina, D. M.; Girolami, G. A. *J. Am. Chem. Soc.* **1988**, *110*, 1643.

15. Itoh, H.; Tsuzuki, Y.; Yogo, T.; Naka, S. *Mater. Res. Bull.* **1987**, *22*, 1259.

16. Grimes, R. N. In *Comprehensive Organometallic Chemistry*; Wilkinson, G.; Stone, F. G. A.; Abel, E., Eds.; Pergamon: Oxford, 1982; Chapter 5.5.

17. Marks, T. J.; Kolb, J. R. *Chem. Rev.* **1977**, *77*, 263.

18. (a) Pierson, H. O.; Randich, E.; Mattox, D. M. J. *J. Less-Common Met.* **1979**, *67*, 381. (b) Pierson, H. O.; Randich, E. *Thin Solid Films* **1978**, *54*, 119. (c) Pierson, H. O.; Mullendore, A. W. *Thin Solid Films* **1982**, *95*, 99. (d) Caputo, A. J.; Lackey, W. J.; Wright, I. G.; Angelini, P. *J. Electrochem. Soc.* **1985**, *132*, 2274. (e) Bouix, J.; Vincent, H.; Boubehira, M.; Viala, J. C. *J. Less-Common. Met.* **1986**, *117*, 83. (f) Takahashi, T.; Kamiya, H. *J. Cryst. Growth* **1974**, *26*, 203. (g) Besmann, T. M.; Spear, K. E. *J. Electrochem. Soc.* **1977**, *124*, 786. (h) Besmann, T. M.; Spear, K. E. *J. Cryst. Growth* **1975**, *31*, 60. (i) Pierson, H. O.; Mullendore, A. W. *Thin Solid Films* **1980**, *72*, 511.

19. Glass, J. S., Jr.; Kher, S. S.; Spencer, J. T. *Thin Solid Films* **1992**, *207*, 15.

20. Glass, J. S., Jr.; Kher, S. S.; Spencer, J. T. *Chem. Mater.* **1992**, *4*, 530.

21. Zange, E. *Chem. Ber.* **1960**, *93*, 652.

22. Shriver, D. F.; Drezdzon, M. A. *The Manipulation of Air-Sensitive Compounds*; Wiley-Interscience: New York, 1986.

23. Kher, S.; Spencer, J. T. *Chem. Mater.* **1992**, *4*, 538.

24. (a) Zhang, Z.; Kim, Y. -G.; Dowben, P. A.; Spencer, J. T. *Mat. Res. Soc. Symp. Proc.* **1989**, *131*, 407-412. (b) Kim, Y. -G.; Dowben, P. A.; Spencer, J. T.; Ramseyer, G. O. *J. Vac. Sci. Technol. A* **1989**, *7*, 2796-2798. (c) Mazurowski, J.; Baral-Tosh, S.; Ramseyer, G. O.; Spencer, J. T.; Kim, Y. -G.; Dowben, P. A. *Mat. Res. Soc. Symp. Proc.* **1991**, *190*, 101-106. (d) Glass, J. A., Jr.; Kher, S.; Hersee, S. D.; Ramseyer, G. O.; Spencer, J. T. *Mat. Res. Soc. Symp. Proc.* **1991**, *204*, 397-402. (e) Glass, J. A., Jr.; Kher, S.; Kim, Y. -G; Dowben, P. A.; Spencer, J. T. *Mat. Res. Soc. Symp. Proc.* **1991**, *204*, 439-444. (f) Kher, S.; Spencer, J. T. *Mat. Res. Soc. Symp. Proc.* **1992**, *250*, 311-316. (g) Hitchcock, A. P.; Wen, A. T.; Lee, S. ; Glass, J. A., Jr.; Spencer, J. T.; Dowben, P. A. *J. Phys. Chem.* **1993**, *97*, 8171-8181. (h) Zych, D.; Patwa, A.; Kher, S. S.; Spencer, J. T.; Kurshner, J.; Boag, N. M.; Dowben, P. A. *J. Appl. Phys.* **1994**, *76*, 3684-3687. (i) Kher, S. S.; Spencer, J. T. *J. Phys. Chem. Solids* **1997**, in press.

25. (a) Kher, S. S.; Tan, Y.; Spencer, J. T. *Appl. Organometal Chem.* **1996**, *9*, 297 (b) Kher, S.; Spencer, J. T. submitted. (c) Kher, S.; Spencer, J. T. *J. Phys. Chem. Solids* **1997**, in press. (d) Glass, J. A., Jr.; Hwang, S.-D.; Datta, S.; Robertson, B.; Spencer, J. T. *J. Phys. Chem. Solids* **1996**, *57*, 563.

26. (a) Wirth, H. E.; Palmer, E. D. *J. Phys. Chem.* **1956**, *60*, 914. (b) Hurd, D. T.; Safford, M. M. U.S. Patent No. 2,588,559, June 26, 1954.

27. (a) Muetterties, E.L. *Boron Hydride Chemistry*; Academic Press: New York, **1975**. (b) Lipscomb, W. N. *Boron Hydrides*; Benjamin: New York, **1963**. (c) Onak, T. In *Comprehensive Organometallic Chemistry*; Wilkinson, G.; Stone, F. G. A.; Abel, E., Eds.; Pergamon: Oxford, **1982**; Chapter 5.4.

Chapter 11

Sol–Gel Processed Materials in the Automotive Industry

Chaitanya K. Narula

Department of Chemistry, Ford Motor Company, P.O. Box 2053, MD 3083, Dearborn, MI 48121

The promise of new applications continues to drive the research on sol-gel processing since it allows for the fabrication of materials in forms such as films, coatings, fibers, foams and powders etc. In this chapter, we will review the automotive applications of sol-gel processed materials in a variety of new devices e.g. alumina based washcoats for three-way catalysts, thin films for electrically heated catalyst devices for exhaust treatment under cold-start conditions, controlled pore size alumina materials for lean burn NOx catalysts, $Pr_xZr_{1-x}O_y$ materials for oxygen storage, and microcalorimeter sensor devices for hydrocarbon sensing. We will present our results on the fabrication and testing of these devices. In addition, we will show that there is a relationship between the precursor, and the resulting crystalline phases and oxygen storage capacity of sol-gel processed $Pr_xZr_{1-x}O_y$ materials. This is demonstrated by employing a mixture of $Pr(O^iPr)_3$ and $Zr(O^iPr)_4$ or a new heterometallic alkoxide, $Pr_2Zr_6(\mu_4\text{-}O)_2(\mu\text{-}OAc)_6(\mu\text{-}O^iPr)_{10}(O^iPr)_{10}$. Furthermore, we will discuss the metastable phases observed on thermal treatment of gels and molecular sieves derived from heterometallic alkoxides of the type $MAl(OR)_x$ where M=La,Ce,Ca,Sr,and Ba, R=^iPr.

The sol-gel process was first employed by Ebelman to fabricate optical lenses in 1846 (1). Since then extensive efforts have been made to prepare materials by this process. The mechanism of gelation has also been determined using silica, alumina and some transition metal oxides as examples. The commercialization of antireflective coatings on architectural glass is one of the success stories of the sol-gel process. Research on sol-gel processed materials continues to enable the fabrication of commercially important materials in various shapes and forms and is summarized in recent volumes (2).

Our group has been actively involved in exploring heterometallic alkoxide precursors for sol-gel processed metal oxides, with a focus on the development of devices for automotive applications. Optical coatings, corrosion resistant coatings, catalyst materials, and sensor devices are among the various applications under investigation. Here, we will review our studies on sol-gel processed materials for automotive catalysts and sensor devices that can be employed to reduce emissions.

Automotive Catalysts and Sensors for Exhaust Treatment

In this section, we briefly describe the automotive exhaust treatment system in order to introduce the reader to the constraints and regulations facing the automotive industry today. A detailed description can be found in ref. 3 and 4. All vehicles sold in the United Stated are equipped with exhaust treatment devices. The automotive exhaust reduction catalyst on gasoline powered vehicles is also called a three way catalyst because it oxidizes hydrocarbons and carbon monoxide and simultaneously reduces NOx. The chemical reactions that take place during exhaust treatment are as follows:

Oxidation

$$2CO + O_2 \longrightarrow 2CO_2$$
$$2HC + 3O_2 \longrightarrow 2CO_2 + 2\,H_2O$$

Reduction/Three-Way

$$2CO + 2NO \longrightarrow 2CO_2 + N_2$$
$$4HC + 10NO \longrightarrow 4CO_2 + 2H_2O + 5N_2$$

The substrate for the catalyst is a honeycomb (Figure 1) made of cordierite with a cell density of 400 cell/inch^2. The walls of the channels are coated with a washcoat based on alumina that contains oxides of barium, lanthanum and cerium. These oxides slow down the degradation of the surface properties of alumina on exposure to catalyst operating conditions. The catalyst operates at 550°C. The ratio (R) of reducing to oxidizing gases in the exhaust stream is maintained at 1.0 for the efficient operation of the catalyst. The efficiency of the catalyst drops drastically as the ratio moves away from the stoichiometric R value. The air to fuel ratio (14.1) is adjusted based on the response of an oxygen sensor in the exhaust stream, which functions to keep the value of R at 1.0. If the exhaust stream is hydrocarbon rich, cerium oxide provides the oxygen during the brief period needed for the oxygen sensor to detect and adjust the air to fuel ratio.

As a result of efficient exhaust treatment systems, modern vehicles meet federal and California emission standards. In order to further reduce emissions and to meet ultra-low emission goals, efforts are in progress to treat exhaust in cold start conditions. Cold start refers to a conditions, upon starting a vehicle, when the catalyst is at ambient temperature. It takes the catalyst about 120 seconds to reach light off temperatures.

Figure 1   Honeycomb Substrate. (Reproduced from reference 4. Copyright 1996 American Chemical Society.)

Light off is defined as the temperature required for the catalyst to convert 50% of the exhaust constituents. About 70% of emissions are released in the first 120 seconds after cold start of a vehicle in a normal driving cycle. Close mounted catalyst and electrically heatable catalyst devices are among several strategies used to treat exhaust after a cold start. We have found that sol-gel processed materials can play an important role in the preparation of washcoat materials for the close mounted catalyst devices. Furthermore, we designed an electrically heatable catalyst device which employs the deposition of sol-gel processed washcoat materials on a substrate coated with a conducting film.

Sol-gel processed materials can also be used for lean burn catalyst applications. The lean burn operation of vehicles refers to the operation of gasoline powered vehicles with air to fuel ratio of about 22. Fuel economy can also be improved by manufacturing vehicles with lean burn engines and compression-ignited diesel engines. This results in a drastic reduction in hydrocarbon and CO emissions and is accompanied by a steep increase in NOx emissions. This exhaust can not be treated with a three way catalyst. The lean burn NOx catalyst formulations are currently limited to zeolites. These catalysts are being developed for the treatment of diesel exhaust. However, the dealumination of zeolites on hydrothermal aging on exposure to lean burn gasoline exhaust renders them unsuitable for practical applications. Our preliminary efforts to develop lean burn NOx catalyst materials are also summarized.

Sol-Gel Processed Materials for Catalysts

Sol-Gel Processed Materials for Catalysts. Catalysts can be rapidly heated to light off temperatures from a cold start if placed close to the engine. This close coupling leads to exposure of the catalyst to higher temperatures than under current catalyst conditions. This cold start exhaust treatment strategy can work if catalytic materials can be tailored to retain the necessary surface properties at high temperatures. In three way catalysts, the surface properties of the alumina washcoat are stabilized by impregnating alumina with the oxides of barium, lanthanum and cerium. We reasoned that a better distribution of lanthanides and rare earths will improve the thermal stability of the surface properties of alumina and prevent the γ- to α-alumina phase transition. We chose heterometallic alkoxides as precursors for several reasons. They can be easily prepared by methods described by Mehrotra et al. (5).

$$MCl_n + KAl(O^iPr)_4 \longrightarrow M[Al(O^iPr)_4]_n + n\,KCl$$

In this case, the chemistry of a mixture of alkoxides and heterometallic alkoxides is not significantly different because heterometallic alkoxides are present in a solution of a mixture of alkoxides of alkaline earths and rare earths and aluminum alkoxides.

$$M(O^iPr)_n + Al(O^iPr)_3 \longrightarrow M[Al(O^iPr)_4]_n$$

$$n = 2, M = Ba$$
$$n = 3, M = La, Ce$$

The advantage of using heterometallic alkoxides of alkaline and rare earths with aluminum is that they can be purified in high yields by sublimation or distillation. The alkoxides of alkaline and rare earths, on the other hand, either do not sublime or sublime in poor yields.

Although the precise mechanism of the hydrolysis of heterometallic alkoxides has not been determined, several groups have found evidence that heterometallic alkoxides do not dissociate in the early stages of hydrolysis (6). An intermediate with a bridging hydroxy group has been isolated and characterized by Caulton et al. supporting the proposed formation of hydroxy bridged intermediates (7).

The gels from heterometallic alkoxides or their mixtures can be prepared by direct hydrolysis (8, 9). The gels derived from $La[AlO^iPr)_4]_3$ remain amorphous below 700°C. Further heat treatment in air at 900°C leads to crystallization of $LaAlO_3$ [average grain size is 15 nm from the Scherrer formula] while alumina remains amorphous. The BET surface area decreases from 80 to 40 m^2/g in the 500-900°C range. Cerium oxide starts to crystallize out on heating the gels derived from $Ce[Al(O^iPr)_4]_3$ at 600°C. The crystallites of CeO_2 increase in size from 2 nm to 15 nm on calcination from 600°C to 900°C. The BET surface area of the 900°C calcined sample was 58 m^2/g. The gels derived from $Ba[Al(O^iPr)_4]_2$ form $BaAl_2O_4$ and $BaAl_2O_4.H_2O$ after heat treatment in air at 400°C. The BET surface area decreases from 80 to 3 m^2/g as the gel is heated from 300°C to 500°C in air.

The gel derived from a mixture of $La[Al(O^iPr)_4]_3$ and $Ce[Al(O^iPr)_4]_3$ in 1:1 ratio shows only a CeO_2 crystalline phase even after calcining at 900°C. High resolution transmission electron microscopy (HREM) and energy dispersive spectroscopy (EDS) shows that crystalline lanthanides concentrate on amorphous alumina particles (Figure 2). The lanthanum to cerium ratio varies from being lanthanum rich to cerium rich. The lattice fringes in all lanthanide areas correspond to (111) CeO_2 only, suggesting that La_2O_3 forms a solid solution with CeO_2.

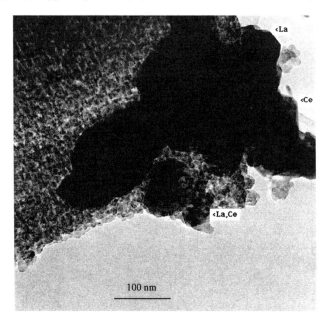

Figure 2   Transmission Electron Micrograph of Gel Derived from a Mixture of $La[Al(O^iPr)_4]_3$ and $Ce[Al(O^iPr)_4]_3$ after Sintering at 1000°C. (Reproduced with permission from reference 8. Copyright 1997 The Royal Society of Chemistry.)

A mixture of $Ba[Al(O^iPr)_4]_2$, $La[Al(O^iPr)_4]_3$, and $Ce[Al(O^iPr)_4]_3$ in 1:1:1 ratio furnishes a gel which has a BET surface area of 300 m^2/g after heating at 500°C. The surface area reduces drastically on calcination at 900°C in air. The X-ray powder diffraction pattern (XRD) shows poorly crystallized CeO_2 after calcination at 700°C. The oxides of barium, lanthanum, and aluminum remain amorphous even after calcining at 900°C.

These results show that a finer distribution of rare and alkaline earths in the alumina matrix leads to materials which retain surface area even after calcination at high temperatures. These materials are being tested as washcoats for close coupled catalysts.

Diesel Exhaust Oxidation Catalyst. Diesel exhaust contains particulate materials, sulfur oxides, water, HCs and NOx. The particulates are made of non-oxidized lube oil soaked on to soot particles. The purpose of an oxidation catalyst in diesel exhaust is to oxidize the lube oil hydrocarbons which do not participate in NOx reduction. Farrauto et al. found that catalyst efficiency and light-off temperatures can be predicted on the basis of normalized DTA areas and onset temperatures, respectively, from a simultaneous TGA/DTA of a lube oil soaked catalyst formulation (*10*). The mechanism appears to involve oil adsorption in the pores of the washcoat.

To address the issue of lube oil hydrocarbon oxidation, we reasoned that sol-gel processed alumina materials incorporating cerium can play a role in diesel catalyst formulations, provided the materials can be made with a controlled pore size and show hydrothermal stability under diesel catalyst operating conditions. We selected BASF Pluronic PL-64 as a template to introduce a controlled pore size distribution with long range crystallographic order. This surfactant is a copolymer of polyethylene oxide and polypropylene oxide and has been previously utilized by Pinnavaia et al. in the preparation of mesoporous alumina molecular sieves (*11*). We found that the gel forms immediately upon addition of water to a solution of $M[Al(O^iPr)_4]_3$, M = La, Ce and Pluronic PL-64 in 2-butanol. The XRD shows a peak at $2\theta = 0.8°$ which is characteristic of ordered alumina molecular sieves. The surface properties of CeO_2-Al_2O_3 sieves [BET surface area 226 m^2/g, average pore diameter 61.7 Å, and BJH desorption pore diameter 44.0 Å] are not significantly different from the CeO_2-Al_2O_3 gels prepared by direct hydrolysis of $Ce[Al(O^iPr)_4]_3$.

The normalized DTA peak area obtained from a simultaneous TGA/DTA of diesel lube oil soaked CeO_2-Al_2O_3 molecular sieves was 30.5 µV.min/mg-sample/mg-lube oil as compared to that for a CeO_2-Al_2O_3 gel at 29.7 µV.min/mg-sample/mg-lube oil. The DTA onset temperature for the CeO_2-Al_2O_3 molecular sieves was 178°C while the onset temperature for the CeO_2-Al_2O_3 gel was 210°C. These results suggest that the crystallographic order of the pore structure of the washcoat can play a role in reducing the light-off temperature but has little impact on diesel oxidation catalyst efficiency (*12*).

Oxygen Storage Materials. Short term oxygen demand can be addressed by oxygen storage materials in the time lag associated with any adjustment of the air/fuel ratio, after the oxygen sensor signal is received by the control module. Thus, three way catalysts that operate in slightly rich or lean conditions contain cerium oxide as a oxygen storage material. The reactions associated with the process are as follows:

Rich conditions

$$2\ CeO_2 + CO \longrightarrow Ce_2O_3 + CO_2$$

Lean conditions

$$Ce_2O_3 + 1/2\ O_2 \longrightarrow 2\ CeO_2$$

As a general rule, an increase in atomic number in the lanthanide series (Ce, Pr, Tb) is accompanied by decreased thermal stability in the corresponding oxides. It follows that the oxygen in PrO_2 should become available at lower temperatures than in CeO_2. However, PrO_2 has not been used as an oxygen storage material because it reacts with the alumina washcoat to form $PrAlO_3$. Praseodymium is stabilized in the +3 oxidation state in $PrAlO_3$ and can not undergo the PrO_2 to Pr_2O_3 transition necessary for oxygen storage. Catalyst formulations based on CeO_2-ZrO_2 have recently been developed which have thermally stable surface areas and can also act as washcoats for palladium. The CeO_2-ZrO_2 system is a single phase solid solution crystallized in a fluorite-like structure. Analogous PrO_2-ZrO_2 materials have not been studied because they are present only as mixtures with primarily perovskites and other phases.

We found that PrO_2-ZrO_2 gels can be prepared from the hydrolysis of a mixture of $Pr(O^iPr)_3$ and $Zr(O^iPr)_4 \cdot {}^iPrOH$ (*13*). The dried and calcined (in air) gels furnish single phase PrO_2-ZrO_2 solid solution materials crystallized in a fluorite-type structure. These materials did not contain other crystalline phase contaminants. Alumina particles can also be impregnated with PrO_2-ZrO_2 sol and subsequently calcined in air to produce PrO_2-ZrO_2 coated alumina particles. A high resolution electron micrograph shows a thin layer of PrO_2-ZrO_2 on the alumina particles (Figure 3). Since PrO_2 remains a part of the fluorite structure with ZrO_2, the oxygen storage capacity is retained.

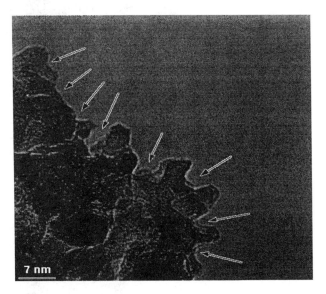

Figure 3 Transmission Electron Micrograph of Alumina Coated with Gel Derived from a Mixture of $Pr(O^iPr)_3$ and $Zr(O^iPr)_4 \cdot {}^iPrOH$ after Sintering at 800°C.

Our attempts to use heterometallic alkoxides to prepare PrO_2-ZrO_2 materials were not successful. The heterometallic alkoxide, $Pr[Zr(O^iPr)_5]_3$, is not a practical precursor because it requires several months to crystallize out $Pr[Zr(O^iPr)_5]_3$ from a mixture of $Pr(O^iPr)_3$ and excess $Zr(O^iPr)_4$ (*14*). We prepared a new heterometallic alkoxide from the reaction of $Pr(OAc)_3$ and $Zr(O^iPr)_4$ in refluxing toluene.

$$2\ Pr(OAc)_3\ +\ 6\ Zr(O^iPr)_4.^iPrOH\ \longrightarrow\ Pr_2Zr_6(\mu_4\text{-}O)_2(\mu\text{-}OAc)_6(\mu\text{-}O^iPr)_{10}(O^iPr)_{10}$$

The alkoxide can be crystallized from its concentrated solution in toluene. The proposed formula for the alkoxide is based on a well characterized gadolinium analogue prepared by the reaction of gadolinium acetate with zirconium 2-propoxide (*15*). Elemental analysis of the alkoxide corresponds to the proposed formula and the infra-red spectra shows peaks for acetate groups and 2-propoxy groups. The gels prepared from this alkoxide furnish PrO_2-ZrO_2 powders with no fluorite phase indicating that this class of heterometallic alkoxides are not suitable precursors.

These results demonstrate that the desired crystalline phase and consequently oxygen storage activity of materials can be tailored by careful selection of the precursors.

Lean Burn NOx Catalyst. Lean burn gasoline and diesel engines run oxygen rich and the current three way catalyst is inefficient for NOx reduction under low hydrocarbon conditions. Furthermore, the exhaust temperature for lean burn gasoline is about 200°C higher than the diesel exhaust which is generally catalyzed at 350°C.

Among the catalysts reported to selectively reduce NOx under lean conditions, zeolite based catalysts have been found to be effective for the treatment of automotive exhaust (*16*). However, zeolite based catalysts are hydrothermally unstable at temperatures close to the upper limits of diesel exhaust catalytic temperatures and fail on extended exposure to exhaust temperatures from lean burn gasoline engines. We found that silver deposited on a commercial alumina washcoat is as effective as Cu-ZSM-5 for NOx reduction at 500°C when tested for a gas mixture containing 1,000 ppm NO, 1500 ppm C_3H_6 and a space velocity of 25,000 hr^{-1}. A 96% conversion of NOx can be achieved if commercial alumina is replaced with sol-gel processed alumina (*17*). This improvement can be attributed to a fine distribution of silver in the pores of sol-gel processed alumina and reduced sintering of the silver particles.

Devices. In addition to washcoat materials described in the previous sections, we have employed sol-gel processed materials to fabricate electrically heatable catalyst devices and microcalorimeter sensors. These devices are described in the following sections.

Electrically Heatable Catalyst. A second strategy to treat exhaust gases during a cold start involves electrically heating the catalyst to light-off temperatures. Commercially available electrically heatable catalyst devices rely on heating the whole catalyst brick to the light-off temperature after a cold start. This process is energy intensive and requires 1.0-1.5 kW of power. The available power from the battery on a vehicle is <1.0 kW. Efforts have been made to reduce the power requirement by reducing the size of the catalyst. This also results in a reduction in surface area available for the treatment of exhaust.

Our design of an electrically heatable catalyst is based on the concept that only the noble metals and gas molecules being treated need to be at the light off temperature (*18, 19*). This can be accomplished by placing a conducting layer on the substrate under the washcoat. In order to test this hypothesis, we fabricated a prototype (Figure 4) on a glass substrate coated with fluoride doped tin oxide.

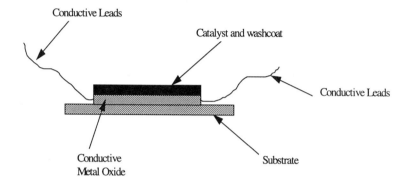

Figure 4   A Prototype for Electrically Heatable Catalysts. (Adapted from reference 4. Copyright 1996 American Chemical Society.)

We connected two electrical contacts on the glass coupon of size 2 cm × 2 cm and coated the coupon with sol-gel processed titania. After firing the film at a rate 2°C/min to 550°C, the film was coated with a PdCl$_2$ solution. The PdCl$_2$ was decomposed by firing the film to 550°C at a rate of 2°C/min. This device was placed in a flow reactor and heated by applying power to the film thorough electrical leads. The light-off performance as a function of power and temperature are shown in Figure 5. From the power requirements of this prototype, it can be estimated that a device based on this concept will require <1 kW to reach light off temperatures, instantaneously.

Microcalorimeter Sensors. In order to meet the new on-board diagnostic requirements mandated by the Clean Air Act, it is necessary to place a sensor on a vehicle to monitor the performance of the catalyst. Microcalorimeter sensors are of interest due to their simplicity, sensitivity and relatively fast response time as measured by change in resistance. This class of sensors measures the heat evolved on the catalytic oxidation of hydrocarbons and CO. Commercially available sensors lack the sensitivity and fast response time necessary to monitor the performance of catalyst in a dynamic exhaust stream by measuring the untreated concentration of hydrocarbon in the emissions.

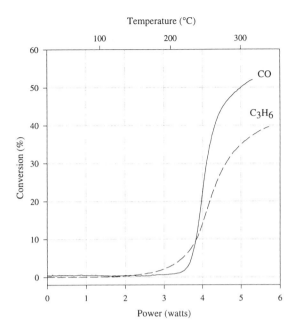

Figure 5   Light-Off Performance of the Prototype for an Electrically Heatable Catalyst. (Reproduced with permission from reference 19. Copyright 1998 American Institute of Chemical Engineers.)

In order to improve the sensitivity of the microcalorimeter sensor, the Physics department at Ford used silicon micromachining to produce a thin membrane sensor to measure the small amounts of heat generated. The sensor (Figure 6) is fabricated as two   2 mm × 2mm reference and active membranes, each 2 micron thick. We deposited a 600Å thick film of alumina on both membranes from an alumina sol obtained by the hydrolysis of 2,4-pentanedione modified aluminum 2-propoxide (*20, 21*). Palladium was deposited on one of the alumina coated membrane (active) from a 30g/L Pd(1,5-diphenyl-1,4-pentadien-3-one)$_2$ solution in THF using a microsyringe. In this arrangement, the temperature difference between the active and reference membranes is due to the heat evolved during the catalytic oxidation of HC and CO on the active membrane.

The device was tested at 400°C in a gas mixture with a 0.5 ratio of reducing to oxidizing gases and a flow rate of 250 sccm. The results (Figure 7) show that the sensor response for propene and CO is linear with a sensitivity of 3°C/1000 ppm and 0.85°C/1000 ppm, respectively.

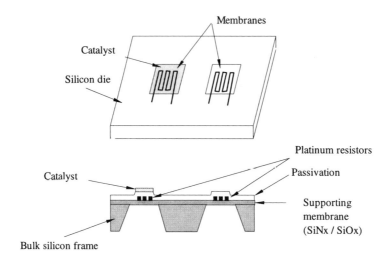

Figure 6   A Microcalorimeter Sensor. (Adapted Courtesy of reference 20.)

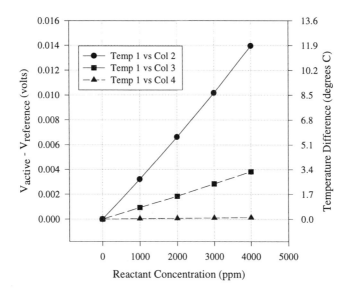

Figure 7   The Response of Microcalorimeter Sensor Operating at 400°C. (Reproduced with permission from reference 21. Copyright 1996 Materials Research Society.)

Conclusions

In conclusion, we have shown that sol-gel processed materials can be employed in preparing metastable phases of washcoat materials. The lanthanide containing alumina materials retain their surface properties even after thermal treatment at 900°C. The sol-gel process also allows for the optimum preparation of PrO_2-ZrO_2 materials in the fluorite phase which is the active phase for oxygen storage. It is also economical to fabricate devices employing sol-gel processed materials. Thus, sol-gel processed materials can find applications in automotive exhaust treatment devices despite the cost of some precursor materials and the loss of some of the beneficial properties on aging at elevated temperatures.

References

1.   Ebelman, J.J. *Ann.* **1846**, *57*, 331.
2.   Brinker, C.J.; Scherer, G.W. *Sol-Gel Science, The Physics and Chemistry of Sol-Gel Processing*, Academic Press: New York, 1990. Klein, L.C. *Sol-Gel Technology for Thin Films, Fibers, Preforms, Electronics, and Specialty Shapes,* Noyes Publications: New Jersey, USA, 1988.
3.   Heck, R.M.; Farrauto, R.J. *Catalytic Air Pollution Control: Commercial Technology*; Van Nostrand Reinhold: New York, 1995.
4.   Narula, C.K.; Allison, J.E.; Bauer, D.R.; Gandhi, H.S. *Chem. Mater.* **1996**, *8*, 984.
5.   Mehrotra, R.C.; Agrawal, M.M.; Mehrotra, A. *Synth. Inorg. Metal-Org. Chem.* **1973**, *3*, 1981.
6.   Campion, J.-F.; Payne, D.A.; Chae, H.K.; Maurin, J.K.; Wilson, S.R. *Inorg. Chem.* **1991**, *30*, 3245. Hubert-Pfalzgraf, L.G. *Polyhedron* **1994**, *13*, 1118.
7.   Kuhlman, R.; Vaarstra, B.A.; Streib, W.B.; Hufman, J.C.; Caulton, K.G. *Inorg. Chem.* **1993**, *32*, 1272.
8.   Narula, C.K.; Weber, W.H.; Ying, J.Y.; Allard, L.F. *J. Mater. Chem.* **1997**, *7*, 1821.
9.   Narula, C.K. *Ceramic Transaction* **1997**, *73*, 15.
10.   Farrauto, R.J.; Voss, K.E. *Applied Catalysts B: Environmental* **1996**, *10*, 29.
11.   Bagshaw, S.A.; Prouzet, F. Pinnavaia, T.J. *Science* **1995**, *269*, 1242.
12.   Narula, C.K.; Shinomiya, H.; Zinbo, M.; Lowe-Ma. C.K.; Plummer Jr., H.K. Unpublished results.
13.   Narula, C.K.; Taylor, K.L.; Haack, L.P.; Allard, L.F.; Datye, A.; Sinev, M. Yu.; Shelef, M.; McCabe, R.W.; Chun, W.; Graham, G.W.; *Mater. Res. Soc. Symp. Proc.*, **1998**, *497*, 15.
14.   Turova, N. Ya; Kozlova, N.I.; Novoselova, A.V. *Russ. J. Inorg. Chem.*, **1980**, *25*, 1788.
15.   Daniele, S.; Hubert-Pfalzgraf, L.G.; Daran, J.C.; Toscano, R.A. *Polyhedron*, **1993**, *12*, 2091.
16.   Shelef, M. *Chem. Rev.*, **1995**, *95*, 209.
17.   Narula, C.K.; Jen, H-W.; Gandhi, H.. U.S. Patent application, 08/311,298, Sept 23, 1994.

18.   Narula, C.K.; Visser, J.H; Adamczyk, A.A. US 5,536,857, 1996.

19.   Nakouzi, S.R.; McBride, J.R.; Nietering, K.E.; Visser, J.H.; Adamczyk, A.A.; Narula, C.K. *AIChe Journal,* **1998**, *44*, 184 .

20.   Visser, J.H.; Narula, C.K.; Zanini-Fisher, M.; Logothetis, E.M.; US Patent 5,707,148, 1998.

21.   Nakouzi, S.R.; McBride, J.R.; Nietering, K.E.; Narula, C.K. *Mater. Res. Soc. Symp. Proc.* **1996**, *431*, 349.

22.   Narula, C.K. et al. *Mater. Res. Soc. Symp. Proc.* **1998**, *487*, 15.

Chapter 12

Covalent Modification of Hydrogen-Terminated Silicon Surfaces

Namyong Y. Kim[1] and Paul E. Laibinis[2]

[1]Departments of Chemistry and [2]Chemical Engineering, Massachusetts Institute of Technology, Cambridge, MA 02139

Alcohols and Grignard reagents react with the hydrogen-terminated surfaces of porous silicon, Si(100), and Si(111) and form covalently attached organic layers. With alcohols, the reaction occurs at temperatures of 40 to 90 °C and is compatible with the presence of functionalities such as halides, olefins, esters, and carboxylic acids within the reacting alcohol; the resulting films attach to the silicon surface by Si-O linkages. With Grignard reagents, the reaction occurs at room temperature and forms Si-C bonds with the support. For both the alcohols and Grignard reagents, their attachment to the surface occurs concurrently with the cleavage of Si-Si bonds and an etching of the silicon framework during the reaction. With Grignard reagents, the level of etching is slight and easily contolled, thereby allowing straightforward, reproducible formation of stable films on the porous and crystalline silicon supports. For both reactions, the organic layer is directly attached to the silicon substrate.

The derivatization of semiconductor surfaces remains as an active area of research due to the technological importance of these materials, the continuing need to produce smaller features with better controlled surfaces and interfaces, and a fundamental interest in the parallels that exist between solution-phase and solid state chemical reactions (*1-13*). Despite the widespread use of silicon in modern electronic devices, few methods are available for chemically modifying its surface beyond those involving vapor deposition (for SiO_2 and Si_3N_4) and oxidation processes that produce thick films. Most solution-phase methods for modifying silicon with molecular films have used various chloro- and alkoxysilanes (*14*); however, these species formally provide attachment to oxides on silicon and not directly to the silicon framework. Direct reaction with a bare silicon surface is complicated by its rapid oxidation; however, recent methods have been reported that modify the unterminated silicon surface with olefins under highly controlled, ultrahigh vacuum conditions (*5-7*).

Our goal was to provide straightforward, solution-phase methods for producing molecular films with direct covalent attachment to the silicon framework by procedures that resembled those used to form self-assembled monolayers (*14*) (i.e., room temperature, solution-phase adsorption at atmospheric pressure). For this procedure, the hydrogen-terminated silicon surface provided a convenient starting material as the etching of the native oxide from crystalline silicon by HF(aq) produces this surface and it provides a temporary passivation for silicon against oxidation at ambient conditions. The hydrogen-terminated Si(100) and Si(111) surfaces are important and frequent intermediates in silicon processing due to their well-defined structure and metastable behavior toward oxidation. Previously reported solution-phase methods for modifying this surface include a photoinitiated reaction with terminal olefins (*2,3*) and a two-step procedure where the surface is first chlorinated radically with PCl_5 and subsequently reacted with a Grignard reagent at elevated temperature (80 °C) for up to 8 days (*4*).

For our investigation, porous silicon provided another convenient starting material that exposes a hydrogen-terminated silicon surface. Porous silicon has been a material of renewed attention since discovery of its luminescent properties in 1990 (*15*). This form of silicon has been suggested for use in silicon-based optoelectronic devices, where porous silicon will likely have an advantage over the use of compound semiconductors because of its compatibility in processing with existing silicon-based fabrication technologies (*16*). Porous silicon is prepared by anodically etching crystalline Si in a HF/ethanol solution to form a thin layer of porous silicon on the silicon support (*15*). For investigating the chemical reactivity of the hydrogen-terminated surface, porous silicon has an advantage over crystalline substrates due to its higher surface area and greater ease of analysis (particularly by infrared spectroscopy). The ability to form covalently attached films on porous silicon would have potential practical implications in that the luminescent properties of porous silicon degrade upon exposure to air. For metal substrates, adsorbed molecular films have been shown to passivate surfaces against oxidation (17) and their attachment to porous silicon could provide possibly similar barrier properties. Reported methods to modify the surface of porous silicon include various photoelectrochemical procedures for the attachment of carboxylates (*10*) and a recent Lewis Acid mediated process using unsaturated hydrocarbons (*13*).

Experimental Section

Single-polished p-Si(100) (1-10 Ω-cm; 50 mm diameter), n-Si(100), and n-Si(111) wafers were obtained from Silicon Sense (Nashua, NH). The p-Si(100) wafers were coated on their unpolished face with an evaporated film of Al (1000 Å) to produce an ohmic contact. Grignard reagents, alcohols, and other reagents were obtained from Aldrich and used as received.

Porous silicon was prepared by anodically etching a p-type Si(100) wafer in 1:1 48% HF(aq)-EtOH in a Teflon cell with a Pt mesh counter electrode. The etch conditions were either 24 mA/cm^2 for 20 min or 10 mA/cm^2 for 5 min followed by 2.5 hr aging in 48% HF solution; these conditions produce a H-terminated layer of porous silicon on the silicon substrate. After etching, the samples were rinsed with EtOH, dried in a stream of N_2, and cut into ~1 x 1 cm^2 pieces. Hydrogen-terminated Si(100) and Si(111) were prepared by etching pre-cut slides in buffered HF for 5 min.

Derivatizations were performed on silicon slides in glass vials under an atmosphere of N_2. With alcohols, the silicon surface was contacted with neat reagent and heated at a specified temperature for reaction; the required reaction time for optimal results was dependent on the alcohol, with typical times being 0.5 to 12 h. With Grignard reagents, the slide was placed in a 1 M solution in THF or ether. After 1-2 h at room temperature, the reaction was quenched by addition of an anhydrous acid (1 M HCl in ether) at room temperature. For both reactions, derivatized samples were rinsed with EtOH and dried thoroughly under a stream of N_2 prior to analysis.

Diffuse reflectance infrared spectroscopy were obtained at a resolution of 2 cm^{-1} using a Bio-Rad FTS 175 spectrometer equipped with an MCT detector and a

Universal Reflectance accessory (4 scans). Attenuated total internal reflectance infrared spectroscopy required use of trapezoidally cut Si(100) and Si(111) crystals with incidence bevels at 45° that were derivatized. X-ray photoelectron spectra were obtained on a Surface Science X-100 spectrometer at a take-off angle of 55° using a monochromatized Al Kα x-ray source and a concentric hemispherical analyzer. Rutherford backscattering spectra were obtained using 2 MeV He^{2+} beam and an energy-sensitive detector positioned at 180° from the incident beam.

Results and Discussion

The covalent modification of hydrogen-terminated crystalline and porous silicon supports was performed using various alcohols, amines, and Grignard reagents. The resulting surfaces were examined by techniques that included infrared spectroscopy (diffuse reflectance for porous silicon and attenuated total internal reflectance for crystalline substrates), x-ray photoelectron spectroscopy (XPS) (for all materials), and Rutherford backscattering (RBS) (for porous silicon). The results in terms of the chemical reactivity and surface yield for crystalline silicon mirrored those obtained on porous silicon, with the latter material providing a greater sensitivity due to its higher surface area. For this reason, we focused experiments on porous silicon and present results here primarily from these experiments; complementary data with crystalline supports provided evidence that the reactions presented here also applied to these substrates, but as they did not provide additional fundamental information, we omit such data from presentation for brevity and to avoid redundancy.

Derivatization of Hydrogen-Terminated Silicon Surfaces with Alcohols. Figure 1 displays diffuse reflectance spectra for porous silicon before and after derivatization. The spectrum in Figure 1a for the as-prepared porous silicon displays characteristic Si-H$_x$ stretching peaks at 2080-2150 cm^{-1} and Si-H$_2$ bending at 914 cm^{-1}, where the former contains three identifiable peaks (2139, 2115, and 2089 cm^{-1}) that can be assigned to silicon tri-, di-, and monohydride species, respectively (*18*). The native porous silicon samples contained oxygenated species as evidenced by the Si-O stretching peak at 1031 cm^{-1}. Exposure of porous silicon to an alcohol (usually neat or as a concentrated solution in tetrahydrofuran or dioxane) at temperatures of 40 to 90 °C resulted in spectroscopic changes to the sample that were indicative of chemical modification.

Figure 1b shows the IR spectrum for porous silicon after derivatization with neat ethyl 6-hydroxy-hexanate [HO(CH$_2$)$_5$CO$_2$CH$_2$CH$_3$]. The spectrum displays peaks at 2850-2960 and 1743 cm^{-1} for the C-H and C-O stretching modes, respectively. Notably, the relative intensity of the O-H stretching absorption at ~3400 cm^{-1} to that for the CO peak in the IR spectrum for the parent compound was greatly diminished in the spectrum of the derivatized sample suggesting that the attachment to the surface occurs with loss of the OH moiety. We infer that the alcohol attaches to the silicon surface by an Si-O bond as evidenced by the increased intensity at ~1080 cm^{-1} for Si-O modes and the appearance of O-Si-H peaks at 2150-2260 cm^{-1}. The spectrum of the derivatized sample exhibited no change after rinsing with various solvents, sonication, or exposure to vacuum, demonstrating that the molecule has been covalently grafted to the silicon surface.

Further proof of the covalent attachment of the compound comes from XPS using both porous silicon and Si(100) as substrates. In Figure 2, the silicon signal for both substrates exhibits the typical doublet for the 2p$_{3/2}$ and 2p$_{1/2}$ emissions, with porous silicon also exhibiting a small amount of oxidized material as noted by peaks at 102-103 eV, compatible with the Si-O peaks observed for the parent material (Figure 1a). In contrast, the spectrum for derivatized Si(100) only displays peaks for Si noting that the derivatization occurs with minimal oxidation of the substrate. The C(1s) region in Figure 2c for derivatized Si(100) displays peaks between 284-290 eV, with a signal at ~289 eV indicative of a highly oxidized carbon species (R$\underline{C}$O$_2$R'), a shoulder at 286-287 eV for less shifted carbon atoms, and a primary peak at ~285 eV

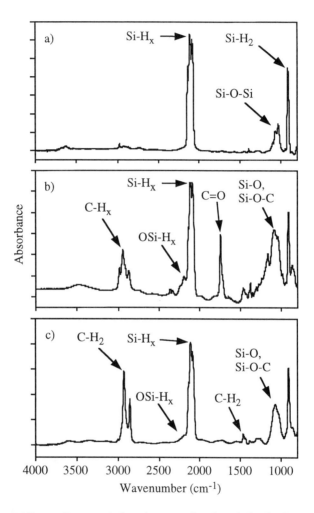

Figure 1. Diffuse reflectance infrared spectra for a) underivatized porous silicon and porous silicon functionalized with b) ethyl 6-hydroxy hexanoate at 87 °C for 20 min, and c) 11-bromoundecanol at 54 °C for 14 h.

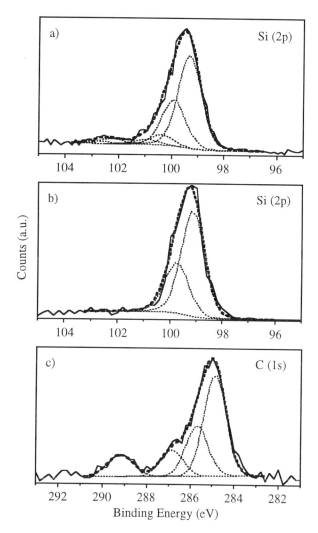

Figure 2. X-ray photoelectron spectra for a) Si(2p) region for porous silicon and b) Si (2p) region and c) C(1s) region for Si(100) derivatized with ethyl-6-hydroxy hexanoate. The dotted lines result from a deconvolution of the spectra (solid line) into their component peaks.

for carbon atoms within a hydrocarbon chain; a spectrum of similar shape was obtained on porous silicon and is not repeated for brevity. The ratio of the different carbon peaks in Figure 2c is compatible with the stoichiometry of the parent compound and confirms its attachment to the surface and the presence of little introduced adventitious carbonaceous material by the procedure. As the conditions for XPS (10^{-9} torr) would cause evaporation from the surface of the parent compound in a physisorbed state, the spectrum also confirms that the reaction produces a robust chemisorbed attachment to the silicon surface. From Figure 2c, we determined that the layer on this flat surface had a surface coverage that was ~90% of that for a densely packed monolayer.

To assess the level of surface modification within the porous silicon framework, we modified its surface by reaction with 11-bromoundecanol [$Br(CH_2)_{11}OH$]. Figure 1c displays the diffuse reflectance infrared spectrum for the derivatized sample. The spectrum displays the presence of the CH stretching modes for the polymethylene chain, where the methyl modes present in Figure 1b are absent. As noted for Figure 1b, the spectrum lacks the O-H stretching peak at ~3400 cm^{-1} that is prominent in the IR spectrum for the parent alcohol. The spectrum in Figure 1c was unchanged after rinsing the sample with various solvents and sonication suggesting covalent attachment to the surface. XPS of the sample (not shown) revealed the presence of bromine that also provided support for the chemical modification. The IR and XPS data confirmed that the chemical reaction provided a chemical modification of the external silicon surface but were unable to establish the level of modification through the porous silicon layer. For our samples, the porous silicon was an outer layer with a thickness of ~10 μm that was supported on a crystalline silicon support. The homogeneity of the derivatization through the porous silicon layer was assess using Rutherford backscattering (RBS) using a 2 MeV He^{2+} source. Figure 3 displays the RBS spectra for a porous silicon sample before and after derivatization with 11-bromoundecanol. For porous silicon, the spectrum shows intensity for He^{2+} particles reflected back from silicon atoms at the surface and deep (~5 μm) into the porous silicon layer. The increasing intensity at lower reflected energies (i.e., at lower channels) probably reflect a a more dense structure to the porous silicon away from the silicon/air interface. After derivatization with 11-bromoundecanol, the presence of bromine is indicated in the RBS spectrum by reflected He^{2+} particles at energies greater than those produced by interaction with the silicon framework. The relatively constant intensity between channels 600 and 800 suggests that the bromine concentration is uniform and present not just at the external surface of the porous silicon but also coats the inner surface of the layer to the depth that RBS can probe in this experiment (~2 μm)

Using this thermally-driven reaction, we have attached both alkyl and aromatic alcohols to the silicon surface covalently and produced films that expose tail groups including CH_3, $CH=CH_2$, CO_2H, CO_2R, and various halogens. The method appears to be general with the exposure time and temperature requiring some fine tuning for the molecule of interest. In all cases, extended exposure resulted in loss of the porous silicon layer and etching of the crystalline substrate; however, covalently attached molecular films could be produced easily by control of the reaction conditions. We note that the derivatization reaction appears to proceed by cleavage of Si-Si bonds to form Si-OR and Si-H bonds upon reaction with the alcohol (ROH) rather than by replacement of Si-H bonds (*11*).

$$H_xSi\text{-}SiH_y + ROH \rightarrow H_xSiOR + SiH_yH \qquad (1)$$

Support for this reaction mechanism (where x and y = 0 to 3 and the remaining coordination sites on silicon are to lattice silicon atoms) is noted by the observation that samples derivatized with deuterium-labelled alcohols (ROD) exhibited Si-D peaks in their IR spectra (*11*).

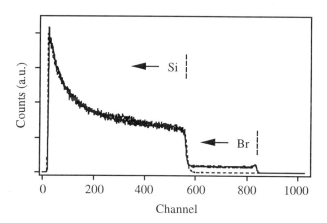

Figure 3.   Rutherford backscattering spectra for underivatized porous silicon (dashed line) and porous silicon derivatized with 11-bromoundecanol at 54 °C for 14 h (solid line).

Derivatization of Hydrogen-Terminated Silicon Surfaces with Amines. The reactions of hydrogen-terminated porous and crystalline silicon with alcohols to form covalently attached films directed the pursuit of chemical methods that might allow similar methods of functionalization of these materials at room temperature without requiring additional thermal input for reaction. We examined the use of alkyl and aromatic amines as reagents on these substrates at room temperature and under conditions up to 60 °C and were unable to obtain spectra (IR or XPS) that indicated their robust attachment to the surface. For porous silicon, we observed a dramatic loss of Si-H intensity by IR along with visual observations that suggested that the amines etch the silicon. The results mirrored those observed with porous silicon when the substrate was exposed to alcohols for extended periods of time; however, we were unable to observe the presence of attached amine species spectroscopically. For both the alcohols and amines, the reactions with the hydrogen-terminated surface appeared to proceed with cleavage of Si-Si bonds that etched the surface

$$H_xSi\text{-}SiH_y + RNH_2 \rightarrow \text{``}H_xSiNHR\text{''} \text{ (not isolated)} + SiH_yH \tag{2}$$

with the difference that generated Si-OR species were more stable and readily isolated than the corresponding Si-NHR species that might be generated by the amines. One hypothesis is that the aminated silicon atoms are more susceptible to further attack by amines and cause the modified silicon atom to be removed from the lattice as soluble $H_xSi(NHR)_{4-x}$ species, thereby leaving behind a hydrogen-terminated silicon surface that shows evidence for being etched but no functionalization.

Derivatization of Hydrogen-Terminated Silicon Surfaces with Grignard Reagents. The discouraging results (in terms of surface modification) using amines were countered by reactions performed using Grignard reagents that should have higher reactivity. Figure 4 displays the IR spectra for porous silicon samples after exposure at room termperature to either an alkyl or aromatic Grignard reagent and a subsequent quench with HCl in ether. After reaction with decylmagnesium bromide, the spectrum in Figure 4a displays C-H stretching frequencies at 2850-2960 cm^{-1} indicative of CH$_2$ and CH$_3$ groups and a CH$_2$ scissor mode at 1464 cm^{-1}. Figure 4b is for a porous silicon sample after reaction with 4-fluorophenylmagnesium bromide and displays the characteristic peaks (1165, 1245, 1498, and 1591 cm^{-1}) for a para-substituted aromatic species. In contrast to the spectra in Figure 1b and c for reactions with alcohols, the SiH stretching modes are similar to those for the original porous silicon (Figure 1a) as they do not display the O-Si-H peaks at 2150-2260 cm^{-1} and the Si-O peak at ~1050 cm^{-1} remains relatively unchanged in intensity. For both samples in Figure 4, the IR spectra were unchanged after rinsing the samples with various solvents, sonication, exposure to vacuum, and notably even after exposure to HF(aq). The spectral evidence suggest that the reaction proceeds by attachment of the Grignard reagent to the silicon support directly as attachments by Si-O-C linkages would yield spectral changes such as those presented in Figure 1 (but not observed in Figure 4) and the attached species that would not survive exposure to HF.

Figure 5 shows XPS spectra for porous silicon before and after derivatization with 4-fluorophenylmagnesium bromide. The survey spectrum for as-prepared porous silicon contains peaks for silicon, oxygen, and adventitious carbon. After reaction with 4-fluorophenylmagnesium bromide, the survey spectrum exhibits a peak for fluorine and additional carbon intensity from the attached fluorophenyl species. Notably, we observed no peaks for magnesium or bromine indicating that the fluorine and carbon peaks are simply due to the presence of unreacted Grignard reagent on the surface. High resolution spectra of the fluorine region revealed the presence of a small amount of fluorine on the as-prepared porous silicon sample, with increased intensity being present after derivatization. The difference in binding energies for the two fluorine species reflects their bonding characteristics. For porous silicon (a), the fluorine is attached to electropositive silicon and appears at a lower binding energy

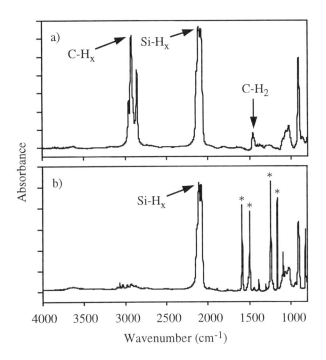

Figure 4. Diffuse reflectance infrared spectra for porous silicon derivatized with a) decylMgBr at room temperature for 2 h and quenched with HCl in ether and b) 4-fluorophenylMgBr at room temperature for 2 h and quenched with HCl in ether. The peaks marked with asterisks (*) in b) are for the 4-fluorophenyl species.

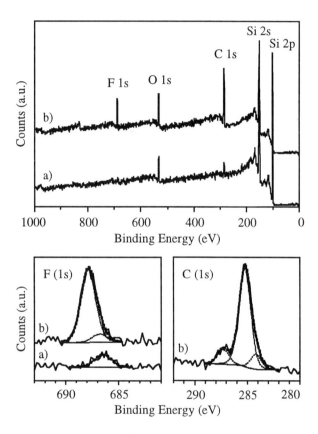

Figure 5. X-ray photoelectron spectra for porous silicon a) before and b) after derivatization with 4-fluorophenylMgBr at room temperature for 2 h and quenched with HCl in ether. The dotted lines result from a deconvolution of the spectra (solid line) into their component peaks.

than for fluorine bonded to carbon in the phenyl ring (b). This shift with bonding to silicon is also observed in the C(1s) spectrum where this region shows a peak at ~287 eV for carbon bonded to an electronegative atom (presumably fluorine), a primary peak at ~285 eV for aromatic carbon (and also adventitous carbonaceous material), and a peak at ~284 eV for carbon bonded to an electropositive atom (presumably silicon). The intensity ratio of the carbon peaks at 284 and 287 eV is ~1:1 suggesting that the presence of each carbon species attached to fluorine requires the presence of a carbon species attached to silicon. At the low pressures of XPS, the volatility of fluorobenzene (the product of 4-fluorophenylmagnesium bromide and HCl) requires that the fluorine signal from the Grignard reagent is from an attached species, where the spectral evidence indicates that the reaction with hydrogen-terminated silicon produces a SiC bond. The observation that the film remains stable during exposure to HF is compatible with the chemical inertness of Si-C bonds to these conditions.

On hydrogen-terminated Si(100) and Si(111) surfaces, similar reactions were observed using Grignard reagents with spectral results that mirrored those presented in Figures 4 and 5. Using XPS on these flat substrates, the surface coverage for reaction with 4-fluorophenylmagnesium bromide was ~50 % of a densely packed monolayer. Extended exposure of substrates to the Grignard reagent did not result in higher surface coverages suggesting that the reaction reaches saturation. Notably, extended exposure of porous silicon to the Grignard did not produce the catastrophic etching noted for the alcohol. The reaction process appears to be similar with the alcohol reaction in terms of its applicability to both porous and crystalline hydrogen-terminated silicon surfaces and to operate by a mechanism that does not involve simple substitution of SiH bonds with Si-OR or Si-C attachments. The reactions differ in that the Grignard reaction proceeds to a maximum level of conversion rather than to continue and etch the substrate as with the alcohols. By analogy to the earlier proposed reactions between the hydrogen-terminated silicon surface and alcohols (equation 1) and amines (equation 2), the Grignard reagent is similarly hypothesized to attach to the surface by breaking lattice Si-Si bonds. This reaction would proceed with formation of a surfacial Grignard species that are quenched during exposure to HCl (or other protic sources). Support for this reaction sequence is noted by reactions using CH_3CO_2D in place of HCl where IR analysis showed the presence of Si-D bonds (*12*); the Si-D peaks were only observed when exposure to CH_3CO_2D was preceded by contact with the Grignard reagent (*12*).

$$H_xSi\text{-}SiH_y + RMgBr \rightarrow H_xSi\text{-}R + SiH_y\text{-}MgBr \qquad (3a)$$
$$SiH_y\text{-}MgBr + HCl \rightarrow SiH_y\text{-}H \qquad (3b)$$
$$SiH_y\text{-}MgBr + CH_3CO_2D \rightarrow SiH_y\text{-}D \qquad (3b')$$

The mechanisms for these reactions (*11,12*) and their continued development are present targets of our work.

Conclusions

Porous and crystalline silicon surfaces bearing a hydrogen termination react with alcohols and Grignard reagents and form covalently attached films. With alcohols, the reaction occurs at temperatures slightly above room temperature and proceeds with concurrent etching of the silicon support, where some level of reaction control is required to functionalize porous silicon surfaces and maintain the structure of the parent substrate. With Grignard reagents, the reaction proceeds at room temperature and is less sensitive to reaction conditions. This latter method forms Si-C bonds to the surface that provide a notable stability to the films to various conditions including exposure to HF. The reactions with alcohols and Grignard reagents apply to both porous and crystalline silicon and provide a general method for attaching organic species covalently and directly to the silicon framework by solution-phase means.

168

Acknowledgments

The authors acknowledge financial support from the Office of Naval Research through their Young Investigator Program.

Literature Cited

1. Sheen, C. W.; Shi, J.-X.; Mårtensson, J.; Parikh, A. N.; Allara, D. L. *J. Am. Chem. Soc.* **1992**, *114*, 1514.
2. Linford, M. R.; Chidsey, C. E. D. *J. Am. Chem. Soc.* **1993**, *115*, 12631.
3. Linford, M. R.; Fenter, P.; Eisenberger, P. M.; Chidsey, C.E. D. *J. Am. Chem. Soc.* **1995**, *117*, 3145.
4. Bansal, A.; Li, X.; Lauermann, I.; Lewis, N. S.; Yi, S. I.; Weinberg, W. H. *J. Am. Chem. Soc.* **1996**, *118*, 7225.
5. Waltenburg, H. N.; Yates, J. T., Jr. *Chem. Rev.* **1995**, *95*, 1589.
6. Hamers, R. J.; Hovis, J. S.; Lee, S.; Liu, H.; Shan, J. *J. Phys. Chem B* **1997**, *101*, 1489.
7. Teplyakov, A. V.; Kong, M. J.; Bent, S. F. *J. Am. Chem. Soc.* **1997**, *119*, 11100.
8. Sturzenegger, M.; Lewis, N. S. *J. Am. Chem. Soc.* **1996**, *118*, 3045.
9. He, J. L.; Lu, Z. H.; Mitchell, S. A.; Wayner, D. D. M. *J. Am. Chem. Soc.* **1998**, *120*, 2660.
10. Sailor, M. J.; Lee, M. J. *Adv. Mater.* **1997**, *9*, 783 and references therein.
11. Kim, N. Y.; Laibinis, P. E. *J. Am. Chem. Soc.* **1997**, *119*, 2297.
12. Kim, N. Y.; Laibinis, P. E. *J. Am. Chem. Soc.* **1998**, *120*, 4516.
13. Buriak, J. M.; Allen, M. J. *J. Am. Chem. Soc.* **1998**, *120*, 1339.
14. Ulman, A. *An Introduction to Ultrathin Organic Films: From Langmuir-Blodgett to Self-Assembly*; Academic Press: San Diego, CA, 1991.
15. Canham, L. T. *Appl. Phys. Lett.* **1990**, *57*, 1046.
16. Hamilton, B. *Semicond. Sci. Technol.* **1995**, *10*. 1187.
17. Laibinis, P. E.; Whitesides, G. M. *J. Am. Chem. Soc.* **1992**, *114*, 9022.
18. Lipp, E. D.; Smith, A. L. In *The Analytical Chemistry of Silicones*; Lipp, E. D., Ed.; Wiley: New York, New York, 1991; Chapter 11.

Chapter 13

Structure, Passivating Properties, and Heterogeneous Exchange of Self-Assembled Monolayers Derived from Chemisorption of the Rigid Rod Arenethiols, $X–C_6H_4–C \equiv C–C_6H_4–C \equiv C–C_6H_4–SH$, on Au(111)

Robert W. Zehner, Richard P. Hsung, and Lawrence R. Sita[1]

Searle Chemistry Laboratory, Department of Chemistry, The University of Chicago, 5735 South Ellis Avenue, Chicago, IL 60637

SAMs derived from the chemisorption of the rigid rod arenethiols **1** and related derivatives are becoming increasingly important for probing conduction at the molecular level. Previous work has shown that the arenethiol **1a** (X = H) adopts a highly ordered $2\sqrt{3} \times \sqrt{3} R30°$ overlayer structure on Au(111), making it the first non-alkane based SAM known to form a commensurate structure. Additional studies described here now address the passivating properties of such arenethiol monolayers as assessed by electrochemical techniques, and their stability towards heterogeneous exchange by hexadecanethiol (HDT). It is concluded that although SAMs derived from **1** appear to possess relatively poor passivating properties when compared to those of SAMs derived from HDT, a process of heterogeneous exchange with HDT can be controlled to "patch" defect sites to provide mixed monolayers that have superior passivating properties and stabilities. The ability to prepare such systems with a variety of ω-functionalized derivatives of **1** facilitates the way for their potential utility in device applications that require control of electrical conduction across interfacial barriers.

Self-assembled monolayers (SAMs) derived from the chemisorption of n-alkanethiols and disulfides on Au(111) now represent one of the most recognized paradigms for order in molecular assemblies (*1-6*). In terms of potential applications, order in SAMs is desirable for promoting the increases in stability and passivation that are critical for corrosion prevention and wear protection, to name just a few, and it is further important for constructing potential electronic and electro-optic devices (*2,7*). As it turns out, however, positional and translational ordering of molecular subunits in SAMs, as opposed to just simple close-packing, is rather difficult to achieve as a number of factors, such as adsorbate structure, adsorbate-surface interactions, adsorbate-adsorbate interactions, and the kinetics of the chemisorption process, are all known to play key roles in dictating monolayer structure. In the case of n-alkanethiol-derived SAMs, it fortuitously happens that the cylindrical 21.5 Å^2 "footprint" size of the all-*trans* n-alkane chains permits formation of a c(4 x 2) superlattice of a commensurate $\sqrt{3} \times \sqrt{3} R30°$ overlayer

---

[1]Corresponding author.

normal in order to reduce free volume and to maximize favorable van der Waals structure in which the chains adopt a tilt angle of roughly 30° off of the surface interactions (5,6). In this almost ideal structural situation, the S⋯S repeat distance is further fixed by the Au(111) lattice constant at 4.995 Å.

In pursuing the design and fabrication of new classes of ordered SAMs on Au(111), commensurate structures are desirable since they should permit the formation of less defective monolayers with larger domain boundaries. Unfortunately, this has proven to be a much more challenging objective than perhaps originally thought since it is now clear that the footprint size and shape of the organothiol adsorbate must meet stringent requirements or else incommensurate structures result (8-10). Indeed, until recently, the only SAMs exhibiting such commensurate ordering were those derived from the original parent n-alkanethiols, $HSCH_2(CH_2)_nCH_3$, and from ω-functionalized n-alkanethiols, $HSCH_2(CH_2)_nCH_2$-X, where X is a small substituent such as OH or NH_2 (6, 11). Based on a proposal originally put forth by Sabatani and co-workers(12), however, we were able to add to this list by preparing and unequivocally demonstrating, through a collaboration with Guyot-Sionnest and Dhirani who performed scanning tunneling microscopy (STM), that SAMs derived from the rigid-rod arenethiol **1** (X = H, **1a**), adopts a highly ordered $2\sqrt{3} \times \sqrt{3}R30°$ overlayer structure on Au(111), making it the first non-alkane based SAM known to form a commensurate structure (13) Further, the STM data for this SAM revealed novel electrical rectifying behavior that appears to be an intrinsic property associated with the molecular and electronic structure of this particular adsorbate (14). Thus, additional investigations of SAMs derived from **1**, and other suitably functionalized derivatives and analogs [e.g., where X = -C≡C-(FeCp_2) (Cp = cyclopentadienyl) (15)] offer a unique opportunity to probe electronic behavior at the molecular and monolayer level (16-19). In this regard, mixed monolayers that incorporate such arenethiol adsorbates and n-alkanethiols are also especially important as they provide a strategy by which to isolate individual molecular conduction pathways for study (16, 20).

1

Given the growing importance of SAMs derived from **1** and related derivatives as vehicles by which new physical phenomena can be discovered and studied, and that can be further, and easily, modified to potentially serve in device applications that require control of electrical conduction across interfacial barriers, a knowledge of the stability and passivating properties of such systems is highly desirable. Accordingly, in this chapter, we discuss the results of electrochemical and exchange studies of SAMs derived from **1** that were conducted in order to address such issues. Part of this work has been previously reported (21).

Results and Discussion

Synthesis of the Arenethiol Adsorbates. The syntheses of a wide range of substituted phenylethynyl oligomers and macrocycles have now been documented, and several different strategies can be utilized for piecing these structures together (15, 22-26). However, in order to minimize the overall number of synthetic steps involved, and to maximize the number of ω-functionalized derivatives of **1** that can be easily accessed, we chose to develop the synthetic route shown in Figure 1. Most importantly, the keys to the success of this route are the simple procedures that can be utilized to prepare large quantities of the two critical building blocks, compounds **2** and **3**. Starting with commercially available 1-bromo-4-

iodobenzene, a one-pot synthesis of the orthogonally protected 1,4-diethynylbenzene derivative **4** can be achieved utilizing the now standard palladium-catalyzed Heck reaction. After purification, the trimethylsilyl group of **4** is then selectively removed to provide compound **2** in a 92% overall yield (*26*). The other building block, 1-iodo-4-thioacetylbenzene (**3**), can similarly be prepared in a straightforward manner by utilizing a one-pot procedure that starts with commercially available 1,4-diiodobenzene as shown in Figure 1 (*27*).

Using the Heck reaction once more, the synthesis of **1a - d** begins with the coupling of a variety of iodobenzenze derivatives (**5a - d**) with compound **2** to provide the "chain-extended" intermediates **4a - d** in excellent yields. The triisopropylsilyl (TIPS) group is then quantitatively removed to provide **7a - d** which are then "end-capped" by coupling with **2** to provide **8a - d**. Finally, the thioester group is removed from these compounds to provide the desired arenethiol derivatives **1a - d** in good overall yield. Significantly, given the concise nature and generally high yields associated with this synthetic route, we strongly believe that a variety of different ω-functionalized derivatives of **1** can now be easily obtained by the non-synthetically inclined investigator who wishes to study the corresponding SAMs. It can finally be mentioned here that this synthesis has recently been extended to additionally provide compounds **1e - g**, where X = NMe$_2$, CH$_3$ and CF$_3$, respectively, however, we have yet to investigate the preparation and properties of SAMs that are derived from them.

Structural Considerations. An essential feature of the adsorbates **1a - g** is that they do not possess alkyl substituents off of the phenyl rings such as those that are commonly employed to enhance the solubility of phenylethynyl oligomers and macrocyclics (*22-26*). The reason for this design feature is that such substituents deleteriously increase the footprint size of the adsorbate, and thereby, eliminate the possibility of forming a commensurate overlayer structure on Au(111). Thus, although such a substituted derivative of **1** has been reported to form a "densely-packed" SAM (*16*), this monolayer most certainly lacks the degree of order and packing density that can be achieved with **1**, and this is an issue when the focus is on SAM stability and passivating properties. In this regard, a fair question to ask is what steric requirements can be placed on the X group in **1** that still permit commensurate ordering? To answer this, we utilize the SAM structural model obtained from the high resolution STM study of the SAM derived from **1a** that is reproduced in Figure 2a (*13*). On the basis of molecular modeling of this structure, we then find that replacement of X = H by X = F, OMe, NO$_2$, NMe$_2$, CH$_3$, and CF$_3$ does not increase the footprint size of **1** enough to disrupt a commensurate packing (e.g., Figures 2b - d). Thus, although it remains to be experimentally verified, we believe that SAMs derived from **1b - 1g** can all adopt a monolayer structure that is similar to **1a**. Finally, a comment can be made regarding the molecular length of these arenethiol adsorbates. From our STM studies, it was noted that adsorbates with shorter chain lengths, such as 4-(phenylethynyl)benzenethiol, form densely-packed SAMs, but ones that lack structural order. We also find that these SAMs have quite inferior stabilities and passivating properties when compared to those derived from **1**. Thus, we believe that the longer molecular lengths in the latter serve to favorably enhance adsorbate-adsorbate interactions that lead to a higher degree of structural order, and consequently, better SAM qualities.

SAM Preparation. Given the poor solubilities of **1a - d** in solvents, such as ethanol, that are commonly used to form SAMs of n-alkanethiols, preparation of SAMs derived from these arenethiols require the use of 1 mM solutions in dichloromethane. Unfortunately, as will be commented on later, we believe that the use of such a solvent reduces the driving force to fill defect sites in the SAM structure with either physisorbed or chemisorbed material, and as a result, poorer

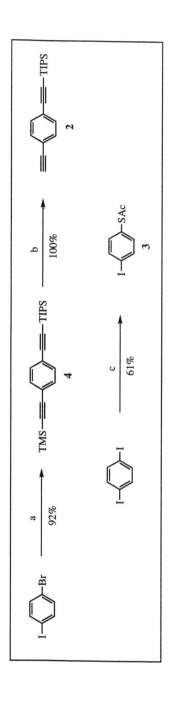

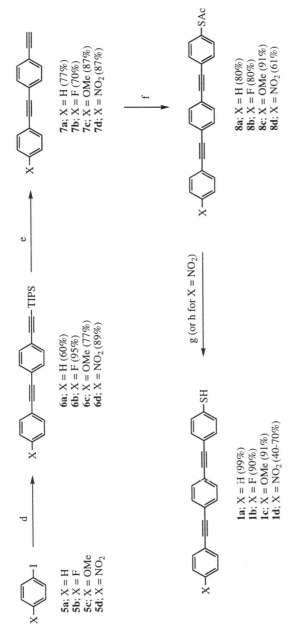

5a; X = H
5b; X = F
5c; X = OMe
5d; X = NO₂

6a; X = H (60%)
6b; X = F (95%)
6c; X = OMe (77%)
6d; X = NO₂ (89%)

7a; X = H (77%)
7b; X = F (70%)
7c; X = OMe (87%)
7d; X = NO₂ (87%)

8a; X = H (80%)
8b; X = F (80%)
8c; X = OMe (91%)
8d; X = NO₂ (61%)

1a; X = H (99%)
1b; X = F (90%)
1c; X = OMe (91%)
1d; X = NO₂ (40-70%)

Figure 1. Synthesis of the arenethiols **1**. (a) (i) 1.1 equiv. triiscpropylsilylacetylene, Pd(PPh₃)₂Cl₂ (4 mol%), CuI (4 mol%), diiscpropylamine, 25° C, 24 h, (ii) 2 equiv. trimethylsilylacetylene, 25° C, 24 h. (b) THF/MeOH, 1 M NaOH (1 equiv.), 25° C, 3 h. (c) (i) 1 equiv. n-BuLi, Et₂O, -78° C, 1 h (ii), S₈ (1.5 eq), -78° C to -10° C, (iii) Ac₂O, imidazole, -10° C to 25° C, 1 h. (d) 1.1 equiv **2**, Pd(PPh₃)₂Cl₂ (4 mol%), CuI (4 mol%), THF/ Hunig's base (1:1), 50° C, 48 h. (e) THF, 1equiv. nBu₄NF (1 M in THF), 30 mir. (f)) 1.1 equiv **3**, Pd(PPh₃)₂Cl₂ (4 mol%), CuI (4 mol%), THF/ Hunig's base (1:1), 50° C, 48 h. (g) (i) 6 equiv LiAlH₄, THF, 25° C, 30 min. (ii) HOAc, Zn, 25° C, 15 min. (h) K₂CO₃, CH₂Cl₂/MeOH (2:1), 20 min.

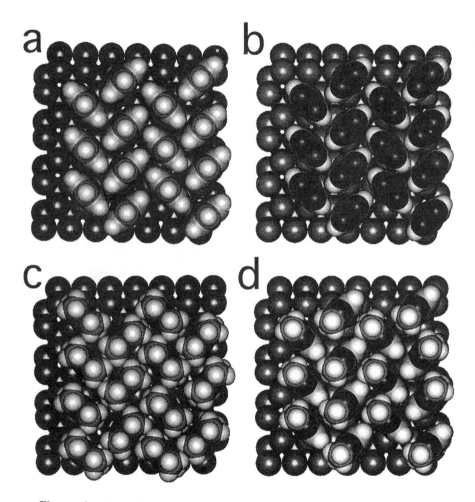

Figure 2. Molecular models showing van der Waals surfaces for SAMs on Au(111) derived from chemisorption of **1** for (a) X = H, (b) X = NO_2, (c) X = NMe_2, and (d) X = OMe.

macroscopic passivating properties are realized when compared alongside SAMs derived from 1 mM solutions of hexadecanethiol (HDT) in ethanol. Accordingly, we find that for different derivatives of **1**, it is best to fine-tune solubility characteristics by using chloroform / methanol solvent mixtures (e.g., 10 : 1 for **1a** and **1b**) to provide SAMs with the best achievable blocking characteristics. Two additional notes can be made regarding preparation of the arenethiol derived SAMs. First, we find, in agreement with McLendon and coworkers (*28*) observations regarding the preparation of SAMs from HDT, that the best SAM passivating properties are obtained when the 150 nm thick polycrystalline Au substrates (prepared by resistive metal evaporation onto crystalline silicon wafers employing a 15 nm Cr adhesion layer) are cleaned by hot acidic peroxide (75% H_2SO_4, 25% H_2O_2) (CAUTION!) for 15 min and then electrochemically cycled (-0.1 to -0.8 V vs SCE in 1 M NaCl) for 5 min. Second, the best SAMs derived from **1**, with respect to film thickness and passivating properties, are obtained after immersion in the substrates for *ca* 18 h. Longer times than this, however, produce turbid solutions as a result of arenethiol oxidation to insoluble disulfide and film thicknesses that are too thick, presumably as the result of a deposition of this material as an overlayer.

Film Thickness and Contact Angles. As can be seen from Table I, SAM film thickness, as measured by ellipsometry, are all consistent with the formation of a single monolayer from **1**. The only exception to this is, in the case not shown, of the nitro derivative **1d** where inconsistent values are obtained that are typically much larger than the calculated value. This inconsistency carries over to other monolayer properties as well, and, to date, we have been unable to reproducibly obtain well-defined SAMs from this particular adsorbate. Although sterically permitted, it may be that electrostatic repulsion between nitro groups prevents formation of a densely-packed SAM structure in this particular system. For the remaining adsorbates, **1a - c**, measured advancing contact angles for water, θ_{adv}, are close to the 80° reported for that of a crystalline surface possessing phenyl rings exposed in an "edge-on" fashion (*29*). Finally, it is interesting to note that, in the case of SAMs derived from **1a** and **1b**, there is a large hysteresis defined by the difference in advancing and receding contact angles, which might be taken as a sign of a poorer SAM homogeneity relative to those derived from **1c** which display a significantly smaller hysteresis (*4*).

Table I. Ellipsometric Film Thicknesses and Contact Angles (θ) (H_2O) for Arenethiol SAMs.

Arenethiol	Film Thickness (Å)[a]	Molecular Length (Å)	θ_{adv} (°)	θ_{rec} (°)
1a	22 ± 4	21	79 ± 3	50 ± 9
1b	22 ± 3	21	71 ± 5	56 ± 6
1c	22 ± 2	23	82 ± 2	75 ± 3

[a]An index of refraction of 1.46 was used for all measurements.
[b]Measured from the van der Waal surface of the sulfur atom to that of the terminal X group.

Electrochemistry. Electrochemical capacitance has proven to be a good measure of the macroscopic permeability of a SAM towards solvated ions, and as such, it can be taken as an indicator of the SAM's integrity (*30-32*). Briefly, if a SAM is impermeable, it should function as an ideal capacitor with the capacitance being given by the Helmholtz relation

$$C = \varepsilon\varepsilon_o/d_{eff}$$

where C is the capacitance per unit area, ε is the dielectric constant of the film, ε_o is the permittivity of free space, and d_{eff} is the SAM film thickness. If the monolayer is impermeable, the charging current observed in the cyclic voltammetry experiment will be independent of potential during the linear potential scans. As a SAM structure becomes more permeable to ions, however, a larger charging current and capacitance will be encountered. Finally, at extreme electrochemical potentials, the large electric field across the SAM can be expected to lead to, at some particular potential value, a failure of the barrier properties of the SAM that is characterized by ion permeation followed by electron-transfer chemistry at the metal-monolayer interface.

Shown in Figure 3 is a comparison of cyclic voltammograms obtained for scans from 0.0 V to +0.15 V and back to 0.0 V vs SCE in 1 M KCl for SAMs derived from **1a** and HDT. As can be seen, the charging current for the former is significantly larger than that of the latter and this correlates with the large difference in their calculated capacitance values (*cf.* $C = 5 \pm 2$ μF•cm^{-2} for SAMs derived from **1a** vs $C = 1.2 \pm 0.1$ μF•cm^{-2} for SAMs derived from HDT). Since both **1a** and HDT are expected to have similar dielectric properties and molecular lengths, their corresponding SAMs should possess nearly identical capacitance values if their barrier properties are the same. The observation that the capacitance for SAMs derived from **1a** is, at a minimum, more than double that for SAMs derived from HDT implies that the former are more permeable, most likely as a result of a higher density of defect sites. The larger variance in the measured capacitance values for SAMs derived from **1a** relative to that for SAMs derived from HDT can then be viewed as a reflection of a greater difficulty in controlling the density of these defect sites from sample to sample. Finally, it can be mentioned that the barrier properties of SAMs derived from **1a** and HDT both fail in 1 M KCl between -0.4 to -0.6 V vs SCE, and this failure causes irreversible damage to the monolayers, as indicated by significantly larger charging currents and capacitance values being observed after this negative scan. On scanning to more extreme positive potentials, the barrier properties of SAMs derived from **1a** fail between +0.5 to +0.7 V in 1 M KCl which is a much lower limit than that obtained for SAMs of HDT (*cf* a failure that occurs at *ca* +1.2 V). However, as can be noted by Figure 4, SAMs derived from **1a** are much more stable to oxidation in nonaqueous electrolytes where higher oxidation potentials can be reached. This increase in stability is seen more clearly in Figure 5 which records measured capacitance as a function of the length of time of repetitive scanning between 0.0 to +0.8 V. As a final note, the capacitance and stability of SAMs derived from **1b** and **1c** are, qualitatively, identical to those of SAMs of **1a**.

The ability of a monolayer to block electron transfer between the gold surface and an electron donor or acceptor in solution has also been shown to be a good measure of its defectiveness (*33-37*). Here, the diffusion-limited currents measured during the cyclic voltammetry experiment at partially blocked electrodes are a function of the distribution of residual pinholes in the monolayer and their size (*38*). In fact, it has been demonstrated that these pinholes can effectively function as a microelectrode array (*39*). Figure 6 shows cyclic voltammograms of monolayer-covered and "bare" gold electrodes in a solution containing 1 mM $K_3Fe(CN)_6$ in 1 M KCl. The voltammogram for bare gold (dashed line) shows the oxidation and reduction peaks typically seen for this redox couple at a metal electrode. On the other hand, the voltammogram recorded using a HDT derived SAM covered gold electrode (dotted line) reveals that this SAM structure efficiently blocks electron transfer, in agreement with previous studies. In contrast to both of

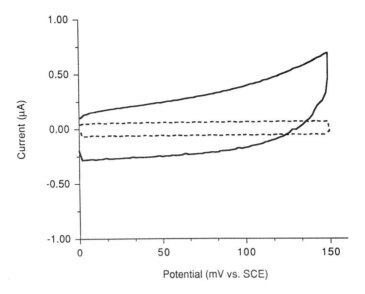

Figure 3. Cyclic voltammograms (first scan) for a gold electrode covered by a monolayer derived from **1a** (solid line) and HDT (dashed line) in aqueous 1 M KCl (electrode area: 0.45 cm^2). (Reproduced with permission from ref. 21. Copyright 1997 American Chemical Society).

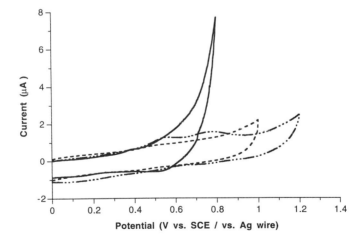

Figure 4. Cyclic voltammograms (first scan) for a gold electrode covered by a monolayer derived from **1a** in aqueous 1 M NaCL (dashed line), 1 mM Bu$_4$NPF$_6$ in THF (solid line), and 1 mM Bu$_4$NPF$_6$ in CH$_2$Cl$_2$ (electrode area: 0.45 cm^2).

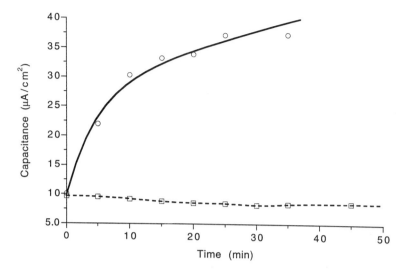

Figure 5. Comparison of the stabilities (as measured by capacitance values) of a monolayer derived from **1a** to continous scanning from 0.0 V to +0.80 V and back to 0.0 V vs SCE in aqueous 1 M NaCL (circles) and 1 mM Bu_4NPF_6 in THF (squares). Lines are aids to the eye only.

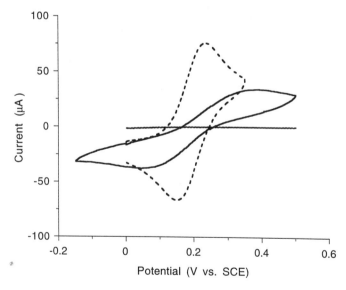

Figure 6. Cyclic voltammograms for a "bare" gold electrode (dashed line) and for gold electrodes covered by a monolayer of **1a** (solid line) and HDT (dotted line) in aqueous 1 M KCl containing 1 mM $K_3Fe(CN)_6$ (electrode area: 0.45 cm^2). (Reproduced with permission from ref. 21. Copyright 1997 American Chemical Society).

these extremes, however, the voltammogram for a gold electrode covered with a SAM derived from **1a** (solid line) indicates that the SAM of this adsorbate apparently has a fairly high defect density that is in keeping with its large capacitance value.

In order to understand the possible origin of the higher defect density in SAMs derived from **1a** - **c** vs those derived from HDT, it is important to comment on the differences in their methods of preparation. As is routinely performed, the HDT SAMs used in this study were prepared from a 1 mM solution of this adsorbate in ethanol. Although this results in a SAM with superior passivating characteristics, recent studies indicate that, due to the poor solvating properties of ethanol for long-chain alkanethiols, the SAM that is obtained is likely composed of a mixture of densely-packed physisorbed and chemisorbed material (*40*). Furthermore, in this case, it is also likely that the strong driving force for adsorption that is provided by this poor solvent is effective in filling defect sites with adsorbed material. A much better solvent for the adsorbate then should lead to a SAM with a higher density of defect sites even though it might contain a higher percentage, or consist exclusively of, chemisorbed material. This idea is supported by our observation of inferior passivating properties, both capacitance and blocking, for SAMs derived from HDT that are prepared from 1 mM solutions of this adsorbate in dichloromethane. A return to the original properties, however, can be achieved by taking this initially prepared sample and soaking it overnight in an ethanolic solution of HDT (1 mM).

In contrast to SAMs derived from HDT, those obtained from **1a** - **c** are prepared using a solution of these adsorbates in dichloromethane. Use of this solvent is mandated by the apparent insolubility of the arenethiols in ethanol. Accordingly, we believe that due to the better solubility of **1** in dichloromethane, there is a large reduction in the driving force to fill defect sites in the SAM structure with either physisorbed or chemisorbed material, and as a result, poorer passivating properties are realized. Another alternative suggestion, however, is that, due to the rigid-rod nature of **1**, the molecular subunits in the SAM derived from these adsorbate are not flexible enough to cover defect sites as well as the much more conformationally mobile HDT molecule potentially can. In this scenario, rigid domain boundaries between ordered assemblies of **1** might then expose the underlying gold surface to a greater extent than in the more fluid HDT SAM structures.

Heterogeneous Exchange. A significant question that is of interest is: what is the stability of SAMs derived from **1** towards heterogeneous exchange with n-alkanethiols? Interestingly, it was found that soaking a gold electrode covered with a SAM derived from **1a** in an ethanolic solution of HDT (1 mM) for 24 h, dramatically improved both the capacitance (cf $C = 2.2 \pm 0.2$ $\mu F \cdot cm^{-2}$) and the blocking characteristics of the monolayer assembly, as well as its stability towards repetitive cycling between 0.0 and +0.8 V (vs SCE) in 1 M KCl (see Figures 7 and 8). Similar improvements in the stabilities and passivating properties of SAMs derived from **1b** and **1c** could be obtained in a likewise manner. Given these results, it was important to determine the composition of the monolayers obtained from this process, and this was achieved by employing a variety of macroscopic characterization techniques. Thus, for the SAM derived from **1a**, ellipsometry provided an apparent film thickness value of 29 ±4 Å after a 24 h treatment that was significantly thicker than that obtained for the pure SAM of **1a** (cf 22 ± 4 Å). However, since repeated washing with both polar and nonpolar solvents had no effect on this measured thickness, the presence of an overlayer of physisorbed material on the SAM can be ruled out. Hence, the increase in the SAM's apparent film thickness is more in keeping with an increase in surface coverage by an adsorbate that most likely arises as a result of a densification of the monolayer that

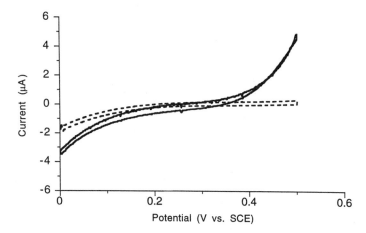

Figure 7. Comparison of the blocking characteristics in 1 M KCl containing 1 mM $K_3Fe(CN)_6$ of a gold electrode covered by a monolayer of HDT (dashed lines) with a gold electrode covered by a monolayer derived from **1a** that has been subsequently immersed in a 1 mM ethanolic solution of HDT for 24 h (solid lines) (electrode area: 0.45 cm^2) (Adapted from ref. 21).

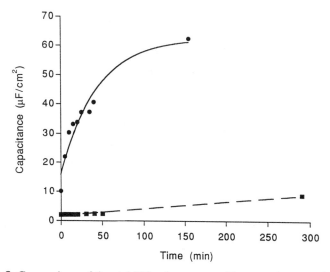

Figure 8. Comparison of the stabilities (as measured by capacitance values) of a monolayer derived from **1a** (filled circles) and a monolayer of **1a** that has been subsequently immersed in a 1 mM ethanolic solution of HDT for 24 h (filled squares) to continous scanning from 0.0 V to +0.80 V and back to 0.0 V vs SCE. Lines are aids to the eye only. (Reproduced with permission from ref. 21. Copyright 1997 American Chemical Society).

occurs with the filling of defect sites (42,43). Advancing contact angles for water, θ_{adv}, also increase from a value of 79 ± 3° for the pure SAM derived from **1a** to a value of 91 ±3° for the patched monolayer after a 24 h treatment indicating that the latter are significantly more hydrophobic, presumably as a result of inclusion of HDT into the monolayer structure. Finally, this increase in hydrophobicity is accompanied by a decrease in the hysteresis defined by the difference between θ_{adv} and the receding contact angles for water, θ_{rec}, which can be taken as further indication of a less defective SAM structure that results upon inclusion of HDT.

In order to examine more closely the composition of the HDT-treated SAMs of **1a**, X-ray photoelectron spectroscopy (XPS) was utilized (44,45). To begin, a series of XPS spectra for a patched monolayer of **1** were recorded at varying takeoff angles to provide a rough depth profile of its composition. The data revealed an increase in intensity of the combined $S(2p_{3/2})$ and $S(2p_{1/2})$ peaks (binding energies: 161.5 and 162.7 eV, respectively) relative to that of the C(1s) peak (BE: 284 eV) as the takeoff angle was increased. This trend supports the notion that the sulfur atoms in the patched monolayer are located beneath the carbon atoms, next to the gold surface, rather than being associated with physisorbed HDT material that is forming an overlayer on the original SAM of **1**. Next, as Figure 9 reveals, a comparison of XPS spectra for pure SAMs derived from HDT and **1a**, and a HDT-treated monolayer of **1a** after a 24 h treatment, exhibit diagnostic differences in the C(1s) peaks that can be used to assess the composition of the latter. More specifically, XPS spectra of SAMs derived from HDT (solid line) and **1a** (dashed line) show a 0.7 eV difference in their C(1s) binding energies (cf BE (HDT): 284.8 eV vs BE (**1**): 284.0 eV) which is in excellent agreement with the values already reported for HDT SAMs(45) and for those of an arenethiol adsorbate that is closely related to **1a**(16). Sandwiched between these two values is the C(1s) peak observed in the XPS spectrum for a SAM derived from **1a** that have been treated with a solution of HDT for 24 h (dotted line). As Figure 9 further indicates, this C(1s) peak can be satisfactorily fit with a linear combination of the two reference spectra for SAMs derived from HDT and **1a**, with the three variable parameters used in this curve-fitting being relative intensity, absolute intensity, and baseline level. Using this analysis, a plot of relative SAM composition as a function of time was obtained that reveals that either a rapid addition of HDT to, or exchange of **1a** by HDT in, the original SAM structure occurs rapidly within the first 30 min. In fact, if this inclusion of HDT were to occur solely by an exchange process, then it would represent replacement of approximately a quarter to a third of the arenethiol adsorbates after the first 24 h. Interestingly, these fractions are in keeping with values obtained from other studies involving the self- and heterogeneous exchange of n-alkanethiol SAMs(46-48). Finally, after this initial period, further slow exchange of **1a** by HDT apparently continues so that after 12 d, only roughly 30 percent of the original arenethiol material in the SAM remains.

A more detailed picture of the exchange process can be obtained from grazing-angle infrared spectra taken at various points in time. By comparing the increase in intensity of the characteristic C-H stretching modes of HDT with the decrease of the phenyl stretching modes for the arenethiol, relative rates for removal of the original monolayer and addition of HDT can be obtained. To measure the loss of the arenethiol monolayer, **1b** was used as the adsorbate since we have observed that the p-fluoro substituent greatly enhances the transition dipole of the longitudinal phenyl stretching mode near 1500 cm^{-1}, thereby, increasing the intensity of the absorption by an order of magnitude relative to SAMs derived from **1a** (21). After 30 minutes of exchange, we observe that the HDT alkane C-H stretching modes observed in the 2800 - 3000 cm^{-1} region, have already grown in significantly, however, they are shifted to higher energy relative to a well-ordered HDT monolayer (see Figure 10a). This shift can be attributed to a disordered, liquid-like state of the alkanethiol adsorbates (31). By comparison, spectra taken

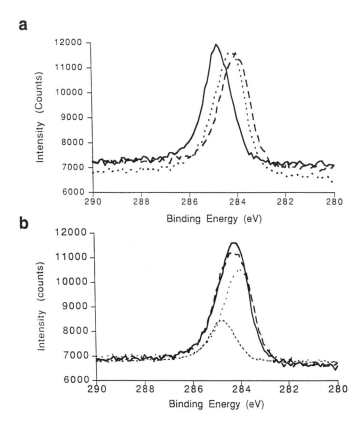

Figure 9. (a) Comparisons of the C(1s) photoelectron peaks from the XPS spectra for a SAM derived from HDT (solid line), a SAM derived from **1a** (dashed line) and a SAM derived from **1a** that has been subsequently immersed in a 1 mM ethanolic solution of HDT for 24 h (dotted line). (b) A best fit (dashed line) for the C(1s) photoelectron peak of a SAM derived from **1a** that has been subsequently immersed in a 1 mM ethanolic solution of HDT for 24 h (solid line) that was constructed from taking linear combinations of the XPS spectra for a SAM derived from HDT (30% relative composition) and a SAM derived from **1a** (70% relative composition). (Reproduced with permission from ref. 21. Copyright 1997 American Chemical Society).

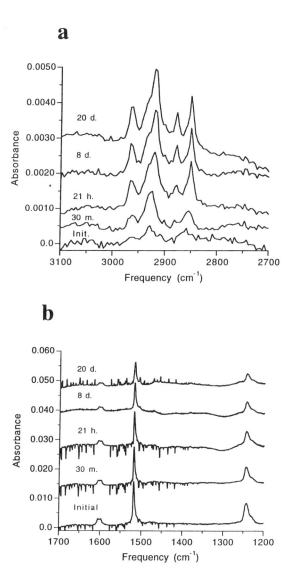

Figure 10. (a) Grazing-angle infrared spectra of (a) the C-H stretching region and (b) of the phenyl stretching region of a monolayer derived from **1b** after (from bottom to top) initial preparation and 33 minutes, 21 hours, 8 days and 20 days immersion in an ethanolic 1 mM HDT solution.

184

after overnight exchange show a smaller amount of additional increase in intensity of these C-H peaks, but now this change is accompanied by a shift to lower energy which is indicative of a quasi-crystalline, well-packed monolayer. Longer exchange times result in a much slower rate of additional HDT incorporation which continues over a period of weeks and results in spectra that resemble even more closely that of a pure HDT-derived monolayer. Finally, as shown by Figure 10b, it is important to note that, during the exchange process, the primary peak in the phenyl stretching region (1518 cm^{-1}) shows a corresponding decrease in intensity with time, however, after overnight exchange, it has only decreased to *ca* 85 percent of its original height, and it remains at 40 - 50 percent after 10 days or more.

Conclusion

Self-assembled monolayers on Au(111) derived from the arenethiols **1** appear to possess relatively poor passivating properties when compared to those of SAMs of HDT. A possible origin for this disparity may lie with the difference in the methods used for the SAM preparations. More specifically, the use of dichloromethane in the former case and ethanol in the latter most likely provides SAMs derived from **1** that have a higher density of vacant defect sites relative to HDT SAMs. In addition, it is possible that the more rigid rod-like nature of the molecular subunits in these SAMs are less effective than the more conformationally flexible n-alkane chains of HDT in providing surface coverage at domain boundaries. However, heterogeneous exchange of SAMs derived from **1** with an ethanolic solution of HDT for short periods of time (e.g. 5 min to 24 h), provides mixed **1**/ HDT monolayers composed primarily of the original arenethiol material, but which now possess superior passivating properties and stability as determined by electrochemical techniques in aqueous electrolytes. Significantly, the exchange process can be attributed to rapid replacement of a small portion of the original monolayer with initially disordered HDT, most likely at defect sites and domain boundaries.

Acknowledgment

This work was supported in part by the MRSEC program under the National Science Foundation (DMR-9400379) for which we are grateful. L. R. S. is a Beckman Young Investigator (1995-1997) and a Camille Dreyfus Teacher-Scholar (1995 - 2000). We wish to also thank Dr. Steven Wasserman at the Argonne National Laboratory for assistance with obtaining and analyzing the XPS data.

Literature Cited

(1) Nuzzo, R. G.; Allara, D. L. *J. Am. Chem. Soc.* **1983**, *105*, 4481-4483.
(2) Swalen, J. D.; Allara, D. L.; Andrade, J. D.; Chandross, E. A.; Garoff, S.; Israelachvili, J.; McCarthy, T. J.; Murray, R.; Pease, R. F.; Rabolt, J. F.; Wynne, K. J.; Yu, H. *Langmuir* **1987**, *3*, 932-950.
(3) Dubois, L. H.; Nuzzo, R. G. *Annu. Rev. Phys. Chem.* **1992**, *43*, 437-463.
(4) Ulman, A. *An Introduction to Ultrathin Organic Films from Langmuir-Blodgett to Self-Assembly*; Academic Press: New York, 1991.
(5) Ulman, A. *Chem. Rev.* **1996**, *96*, 1533-1554.
(6) Poirier, G. E. *Chem. Rev.* **1997**, *97*, 1117-1127.
(7) Li, D. Q.; Ratner, M. A.; Marks, T. J.; Zhang, C. H.; Yang, J.; Wong, G. K. *J. Am. Chem. Soc.* **1990**, *112*, 7389-7390.
(8) Liu, G.-Y.; Fenter, P.; Chidsey, C. E. D.; Ogletree, D. F.; Eisenberger, P.; Salmeron, M. *J. Chem. Phys.* **1994**, 4301-4306.

(9) Wolf, H.; Ringsdorf, H.; Delamarche, E.; Takami, T.; Kang, H.; Michel, B.; Gerber, C.; Jaschke, M.; Butt, H. J.; Bamberg, E. *J. Phys. Chem.* **1995**, *99*, 7102-7107.

(10) Caldwell, W. B.; Campbell, D. J.; Chen, K. M.; Herr, B. R.; Mirkin, C. A.; Malik, A.; Durbin, M. K.; Dutta, P.; Huang, K. G. *J. Am. Chem. Soc.* **1995**, *117*, 6071-6082.

(11) Poirier, G. E.; Pylant, E. D.; White, J. M. *J. Chem. Phys.* **1996**, *105*, 2089-2092.

(12) Sabatani, E.; Cohen-Boulakia, J.; Bruening, M.; Rubinstein, I. *Langmuir* **1993**, *9*, 2974-2981.

(13) Dhirani, A. A.; Zehner, R. W.; Hsung, R. P.; Guyot-Sionnest, P.; Sita, L. R. *J. Am. Chem. Soc.* **1996**, *118*, 3319-3320.

(14) Dhirani, A.; Lin, P.-H.; Guyot-Sionnest, P.; Zehner, R. W.; Sita, L. R. *J. Chem. Phys.* **1997**, *106*, 5249-5253.

(15) Hsung, R. P.; Chidsey, C. E. D.; Sita, L. R. *Organometallics* **1995**, *14*, 4808-4815.

(16) Tour, J. M.; Jones, L., II; Pearson, D. L.; Lamba, J. J. S.; Burgin, T. P.; Whitesides, G. M.; Allara, D. L.; Parikh, A. N.; Atre, S. V. *J. Am. Chem. Soc.* **1995**, *117*, 9529-9534.

(17) Bumm, L. A.; Arnold, J. J.; Cygan, M. T.; Dunbar, T. D.; Burgin, T. P.; Jones, L.; Allara, D. L.; Tour, J. M.; Weiss, P. S. *Science* **1996**, *271*, 1705-1707.

(18) Andres, R. P.; Bein, T.; Dorogi, M.; Feng, S.; Henderson, J. I.; Kubiak, C. P.; Mahoney, W.; Osifchin, R. G.; Reifenberger, R. *Science* **1996**, *272*, 1323-1325.

(19) Samanta, M. P.; Tian, W.; Datta, S.; Hendersen, J. I.; Kubiak, C. P. *Phys. Rev. B.* **1996**, *53*, R7626-R7629.

(20) Sachs, S. B.; Dudek, S. P.; Hsung, R. P.; Sita, L. R.; Smalley, J. F.; Newton, M. D.; Feldberg, S. W.; Chidsey, C. E. D. *J. Am. Chem. Soc.* **1997**, *119*, 10563-10564.

(21) Zehner, R. W.; Sita, L. R. *Langmuir* **1997**, *13*, 2973-2979.

(22) Zhang, J. S.; Moore, J. S.; Xu, Z. F.; Aguirre, R. A. *J. Am. Chem. Soc.* **1992**, *114*, 2273-2274.

(23) Zhang, J. S.; Pesak, D. J.; Ludwick, J. L.; Moore, J. S. *J. Am. Chem. Soc.* **1994**, *116*, 4227-4239.

(24) Ziener, U.; Godt, A. *J. Org. Chem.* **1997**, *62*, 6137-6143.

(25) Jones, L., II; Schumm, J. S.; Tour, J. M. *J. Org. Chem.* **1997**, *62*, 1388-1410.

(26) Höger, S.; Enkelmann, V. *Angew. Chem. Int. Ed., Engl.* **1996**, *34*, 2713-2716.

(27) Pearson, D. L.; Tour, J. M. *J. Org. Chem.* **1997**, *62*, 1376-1387.

(28) Guo, L.-H.; Facci, J. S.; McLendon, G.; Mosher, R. *Langmuir* **1994**, *10*, 4588-4593.

(29) Fox, H. W.; Hare, E. F.; Zisman, W. A. *J. Colloid Sci.* **1953**, *8*, 194.

(30) Finklea, H. O.; Avery, S.; Lynch, M.; Furtsch, T. *Langmuir* **1987**, *3*, 409-413.

(31) Porter, M. D.; Bright, T. B.; Allara, D. L.; Chidsey, C. E. D. *J. Am. Chem. Soc.* **1987**, *109*, 3559-3568.

(32) Chidsey, C. E. D.; Loiacono, D. N. *Langmuir* **1990**, *6*, 682-691.

(33) Finklea, H. O.; Snider, D. A.; Fedyk, J. *Langmuir* **1990**, *6*, 371-376.

(34) Kwan, W. S. V.; Atanasoska, L.; Miller, L. L. *Langmuir* **1991**, *7*, 1419-1425.

(35) Bilewicz, R.; Majda, M. *Langmuir* **1991**, *7*, 2974-2802.

(36) Groat, K. A.; Creager, S. E. *Langmuir* **1993**, *9*, 3668-3675.

(37) Badia, A.; Back, R.; Lennox, R. B. *Angew. Chem. Int. Ed. Engl.* **1994**, *33*, 2332-2335.
(38) Amatore, C.; Savéant, J. M.; Tessier, D. *J. Electroanal. Chem.* **1983**, *147*, 39.
(39) Finklea, H. O.; Snider, D. A.; Fedyk, J.; Sabatani, E.; Gafni, Y.; Rubinstein, I. *Langmuir* **1993**, *9*, 3660-3667.
(40) Castner, D. G.; Hinds, K.; Grainger, D. W. *Langmuir* **1996**, *12*, 5083-5086.
(41) Yang, Z. P.; Engquist, I.; Kauffmann, J.-M.; Liedberg, B. *Langmuir* **1996**, *12*, 1704-1707.
(42) Azzam, R. M. A.; Bashara, N. M. *Ellipsometry and Polarized Light*; North-Holland: New York, 1987.
(43) Smith, T. *J. Opt. Soc. Am.* **1968**, 1069.
(44) Watts, J. F. *An Introduction on Surface Analysis by Electron Spectroscopy*; Oxford University Press: Oxford, 1990.
(45) Bain, C. D.; Troughton, E. B.; Tao, Y.-T.; Evall, J.; Whitesides, G. M.; Nuzzo, R. G. *J. Am. Chem. Soc.* **1989**, *111*, 321-335.
(46) Collard, D. M.; Fox, M. A. *Langmuir* **1991**, *7*, 1192-1197.
(47) Schlenoff, J. B.; Ly, H. *J. Am. Chem. Soc.* **1995**, *117*, 12528-12536.
(48) Peterlinz, K. A.; Georgiadis, R. *Langmuir* **1996**, *12*, 4731-4740.

Chapter 14

New Functionalizable Surfaces Based on Aldehyde-Terminated Monolayers on Gold

Ronald C. Horton, Jr.[1], Tonya M. Herne[2], and David C. Myles[1]

[1]Department of Chemistry and Biochemistry, University of California at Los Angeles, Los Angeles, CA 90095–1569
[2]Chemical Science and Technology Laboratory, National Institute of Standards and Technology, Gaithersburg, MD 20899–0001

2-Hydroxypentamethylene sulfide, a compound easily formed in two steps from readily available materials, forms a novel aldehyde-terminated self-assembled monolayer (SAM) when exposed to a clean gold surface. This aldehyde-terminated SAM then can be used to anchor covalently amine-terminated molecules to the gold surface under mild conditions, through the formation of imines. The immobilization of amines from solution onto the surface has been confirmed through the use of grazing-angle Fourier-transform infrared spectroscopy, X-ray photoelectron spectroscopy, and contact angle measurements. This surface can potentially be used to tether a wide variety of amine-containing molecules to monolayer surfaces.

We present a novel aldehyde-terminated monolayer (1) that, through the formation of an imine bond, can covalently immobilize alkyl amines from solution using no other reagents. Since a wide variety of biomolecules and other amine-containing compounds are available, we see this aldehyde-terminated self-assembled monolayer (SAM) as a promising means of rapidly assembling a wide range of surface-bound systems for many applications (2). We have developed conditions for the formation of the derivatized monolayers and have studied their stability.

Since their initial discovery, SAMs of thiols on gold have sparked considerable interest in the research community (3-9). The ease of design and fabrication of SAMs has led to their application to a wide variety of systems, including micro- and nanofabrication (10-11), prevention of corrosion (12), the determination of reaction mechanisms (13) and modelling of biomembranes (14). Among the most intriguing of these applications is the use of SAMs as biological sensors (15-24). Progress in almost all of these applications, though, is hampered by the need to modify existing compounds or biomolecules with thiol functionalities to allow for monolayer formation. A general foundation monolayer to which a wider variety of compounds can be anchored would greatly facilitate research in these and other areas.

Synthesis of the Aldehyde-Terminated Thiol

2-Hydroxypentamethylene sulfide exists in both open and closed chain forms in solution (Figure 1). This equilibrium is confirmed by the CD_3OD proton NMR spectra, which shows both open and closed forms. In addition, the ^1H NMR reveals the presence of dimeric and oligomeric species. This is believed to be hemithioacetal-linked oligomers, but since the formation of hemithioacetals is highly reversible, we believe this does not result in any hemithioacetal-linked multilayers. Using this equilibrium, the open chain aldehyde-terminated thiol is adsorbed selectively onto the gold surface, generating the desired aldehyde SAM (25). Once this aldehyde-terminated SAM is formed, it then is used to anchor a variety of amine-containing molecules to the gold surface such as alkylamines, enzymes electrochemically active functional groups, chromophores, etc. (Figure 1).

The synthesis of HPMS was carried out in two steps (Figure 2). Pentamethylene sulfide was oxidized with sodium periodate to form the corresponding sulfoxide (26). The sulfoxide was subsequently reacted with trifluoroacetic anhydride, resulting in Pummerer rearrangement (27). Hydrolytic work-up then gave HPMS. Monolayers were formed by immersing clean gold surfaces to a 0.01M solution of HPMS in ethanol under argon for 3-12 hours. After adsorption, the surfaces were rinsed thoroughly with ethanol and dried under a stream of dry argon.

Generation and Derivatization of the Aldehyde-Terminated Monolayer

The aldehyde-terminated SAMs were analyzed using grazing-angle Fourier transform infrared (FTIR) spectroscopy. This spectrum, shown in Figure 3a, clearly shows the three resonances characteristic of an aldehyde. The carbonyl C=O stretching mode appears at 1734 cm^{-1}, typical for alkyl aldehydes (28). Even more indicative is the presence of the doublet at 2724 and 2824 cm^{-1}. These absorbances arise from the Fermi resonance between the aldehydic C-H stretching vibration and the first overtone of the C-H rocking vibration. No strong absorbance due to CH_2 stretching vibration appears. This may be a result of disorder in short-chain monolayer surfaces, leading to broadening of the signal.

The aldehyde-terminated SAM reacts rapidly with alkyl amines in alcohol solvent. Figure 3b shows an FTIR spectrum of an aldehyde-terminated monolayer that has been exposed to a 0.1M solution of dodecylamine in ethanol. The new, derivatized monolayer shows none of the absorbances characteristic of the parent aldehyde-terminated monolayer; the aldehyde Fermi doublet at 2724 and 2824 cm^{-1} and the C=O stretch at 1734 cm^{-1} have completely disappeared. In their place appear strong C-H stretching absorbances at 2856, 2927, and 2967 cm^{-1}, characteristic of a symmetric CH_2 stretch, an asymmetric CH_2 stretch, and an asymmetric CH_3 stretch, respectively. In addition, a strong C=N stretching band is observed at 1670 cm^{-1}. This final absorbance confirms our hypothesis that the aldehyde-terminated monolayer is reacting with the alkyl amine rather than being displaced by it. The absence of a signal at 1670 cm^{-1} in Figure 3c confirms this. The complete disappearance of the aldehyde signals is strong evidence that the condensation reaction has gone very nearly to completion. This is facilitated by the dehydrating effect of the ethanolic solvent, driving the immobilization to completion. We estimate that the detection limit for the aldehyde group stretches is less than 5% of the pure aldehyde-terminated monolayer.

Further proof of the condensation reaction is evinced by exposing the aldehyde-terminated SAM to a series of ethanolic solutions of alkyl amines of varying chain length ($C_nH_{2n+2}NH_2$, abbreviated C_nNH_2). The alkane stretching region in the FTIR spectra of these samples (Figure 4) shows an increase in intensity with

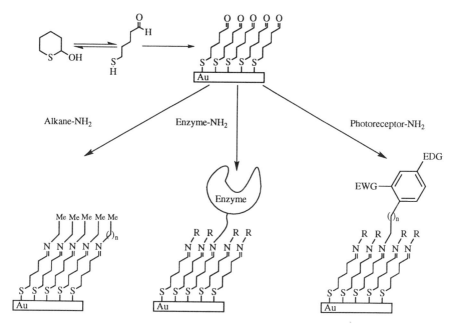

Figure 1. In ethanolic solutions, HPMS is in equilibrium with its straight chain, aldehyde-terminated isomer. The resulting aldehyde-terminated monolayer can then be used to tether a variety of amine-containing compounds to the surface through the formation of imine bonds.

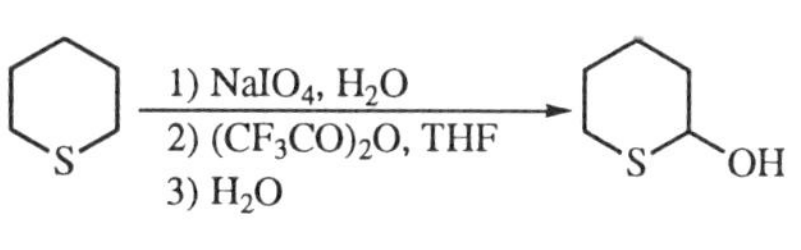

1) $NaIO_4$, H_2O
2) $(CF_3CO)_2O$, THF
3) H_2O

Figure 2. The synthesis of HPMS from pentamethylene sulfide.

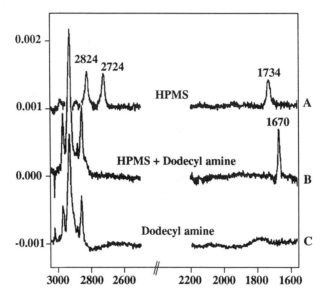

Figure 3. Grazing angle FTIR spectra of (a) aldehyde-terminated monolayer on gold formed from ethanolic solutions of HPMS (b) an aldehyde-terminated monolayer exposed to an ethanolic solution of dodecyl amine (c) a monolayer formed by exposing bare gold to a solution of dodecyl amine.

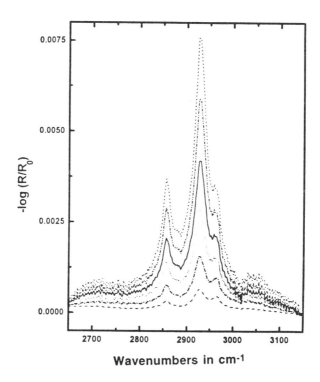

Figure 4. Grazing angle FTIR spectra of the aldehyde terminated monolayer exposed to (– – – –) $C_{10}NH_2$;; (– – – -) $C_{11}NH_2$; (·········) $C_{12}NH_2$; (———) $C_{13}NH_2$; (- - - - -) $C_{15}NH_2$; and (--------) $C_{16}NH_2$. The spectra are stacked in the plot for clarity.

increasing n. The wetting properties of these derivatized SAMs were also examined. The effect of increasing chain length of the alkyl amine on the contact angle with water (Figure 5) causes the surface to increase in hydrophobicity (29). This trend is consistent with a monolayer structure that has the polar imine functionality buried within the monolayer, where it cannot alter the wetting properties of the layer. In addition, the hysteresis of the monolayer decreases with increasing chain length of alkyl amine, indicative of increasing order in the system (9,30).

To investigate further the reaction of alkyl amines with the aldehyde-terminated SAM, we analyzed the derivatized monolayers using XPS. After exposure of the aldehyde SAM to an aqueous solution of dodecyl amine, the XPS spectra shows the appearance of a signal corresponding to a N 1s photoelectron. With increasing chain length of alkyl amines, the Au 4f signal decreases, due to the attenuation by the SAM overlayer. In their study on alkanethiol monolayer systems using XPS, Bain and Whitesides derived the expression: $\ln(Au_n) = -nd/\lambda(\sin\theta) +$ constant (where Au_n = Au 4f intensity for a self-assembled monolayer with n carbons, d = thickness of a single CH_2 {calculated to be 1.1 Å}, λ = escape depth of the Au 4f photoelectron, and θ = take-off angle) (31). They then used this equation to determine the escape depth of photoelectrons from their monolayers. Applying their reasoning to the amine-derivatized aldehyde SAMs system and plotting of ln[Au signal intensity] versus chain length of alkyl amine (Figure 6) using the above expression and assuming d=1.1, we determined that the escape depth of our alkylamine derivatized monolayers is 44 ± 2 Å. This is in close agreement with the value of 42 ± 1Å calculated by Bain and Whitesides indicating that the degree of coverage in the two systems is similar.

Stability and Possible Polymerization of the Monolayer

To determine the stability of the amine derivatized aldehyde SAMs, we monitored their aging properties in air. The wetting properties of the dodecylamine derivatized layer showed no change, even after prolonged (>24 hours) exposure to air. However, aspects of the FTIR spectrum did show time dependence. Although the alkane stretching region is unaffected, the C=N stretching absorbance shows a marked time dependence (Figure 7). After 24 hours of aging in ambient air conditions, the integral of this absorbance decreases to less than 20% of its original value. In analogy to solution chemistry, (Figure 8) the imine bond may succumb to hydrolysis (32), oxidation (33,34), or tautomerization. Hydrolysis seems unlikely as no aldehyde resonances reappear in the FTIR spectra after aging. Oxidation of the C=N bond by ambient O_2 would result in either an oxaziridine or a nitrone. This process also seems unlikely as these transformations usually require stronger oxidizing reagents, such as peroxyacids. A third possibility is tautomerization of the imine to an enamine under acidic or basic catalysis (35). The absence of C=C bond stretching in the FTIR argue against this explanation. It is known that imines can dimerize and polymerize in solution via first tautomerizing to an enamine, then attacking nucleophilically a second imine, and so on (36). We are examining these imine containing SAMs to determine the exact cause of the disappearance of the C=N absorbance.

Summary

In conclusion, we have presented a novel system by which alkyl amines can be attached covalently to a monolayer surface. This aldehyde-terminated SAM has the potential to tether a wide variety of functionalized molecules, thereby encouraging fabrication and application of novel SAM-based sensor systems. The covalent attachment process also possibly has the intriguing aspect of causing polymerization

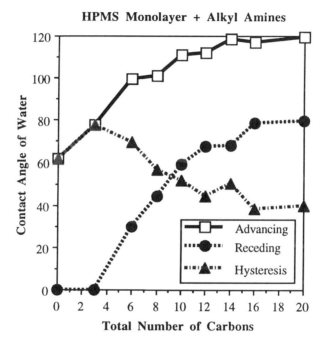

Figure 5. Advancing and receding contact angles for aldehyde-terminated monolayer derivatized with alkyl amines $[(CH3(CH_2)_nNH_2]$, where m = total # of carbon atoms in the monolayer (including aldehyde portion). As longer alkyl amines are immobilized on the aldehyde-terminated monolayer, the advancing contact angle increases, and hysteresis decreases, indicating greater order in the monolayer.

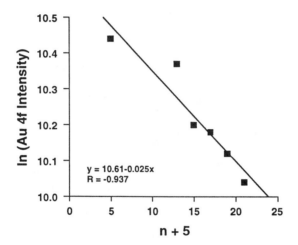

Figure 6. The graph shows the XPS ln(Au 4F intesity) of an aldehyde-terminated SAM derivatized with an n-carbon alkyl amine plotted against the number of carbon atoms in the monolayer (n + 5). The intensity of the Au 4f signal (80-92 eV) at a take off angle of 90°·was determined by measuring the integrated area of the XPS signal.

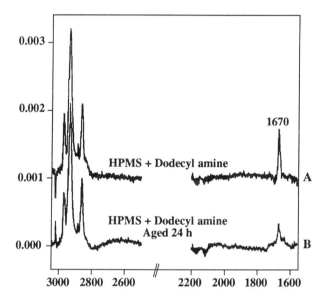

Figure 7. Grazing angle FTIR spectra of (a) an HPMS monolayer derivatized with dodecyl amine and (b) the same monolayer after being exposed to ambient air for 24 hours.

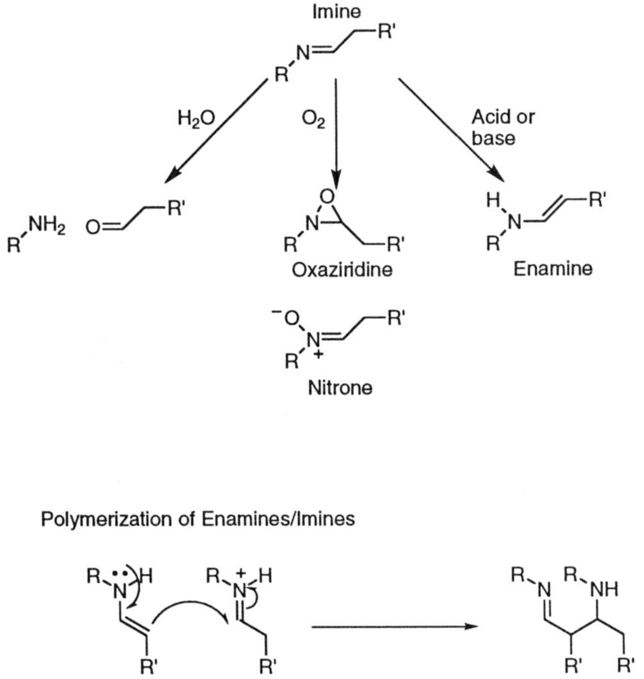

Figure 8. Possible reactions of imines in solution. The imine bond can hydrolyze in the presence of water to generate an amine and an aldehyde. The amine can be oxidized to an oxaziridine or a nitrone. Or the double bond can tautomerize to form an enamine. A final possibility is tautomerization upon formation of the enamine, followed by nucleophilic attack on another imine, causing dimerization and polymerization.

of the base monolayer, thereby increasing the stability of the resulting system. Further work on developing applications of this system and investigating the possible polymerization are being vigorously pursued in our laboratory.

Experimental

Instruments and equipment: FTIR spectra were obtained using an ATI-Matteson Research Series FTIR equipped with a narrow band, liquid nitrogen cooled mercury cadmium telluride (MCT) detector and a Harrick grazing angle accessory. The angle of incidence of the p-polarized light is 74 degrees. 1000 scans at 4 cm^{-1} resolution were collected. The XPS spectra were obtained using the experimental apparatus previously described (37). X-rays were generated by an Al anode operated at 240W. The hemispherical analyzer was operated in the retarding mode at a band pass energy of 20 eV, with the emitted photons collected at 90 degrees with respect to the surface normal. Certain commercial products and instruments are identified to adequately specify the experimental procedure. In no case does such identification imply endorsement by the National Institute of Standards and Technology.

Materials: Pentamethylene sulfide is available from Aldrich Chemical Co., 1001 West Saint Paul Avenue, Milwaukee, Wi 53233. Gold surfaces were prepared by evaporating 200Å of chromium, then 2000Å of gold at 4×10^{-6} torr onto a polished silicon wafer that had been cleaned with piranha etch solution (3:1 conc. H_2SO_4/30% H_2O_2). CAUTION: This solution reacts violently with organic compounds and should be used with extreme caution.

Derivatization of Aldehyde-Terminated Monolayer:
A general procedure for the derivatization of aldehyde-terminated SAMs by alkyl amines is as follows: the aldehyde SAM was placed in a 0.1M ethanolic solution of the alkyl amine under argon for at least three hours. After exposure, the layer was rinsed thoroughly with ethanol and dried under a stream of argon.

Acknowledgments. The NSF (CHE 9501728) provided major funding for this research. Additional funds were obtained from the Academic Senate of UCLA, the Office of the Chancellor, the National Institute of Health (GM08496), and the Rohm and Haas Company.

Literature Cited

1.   Horton, R. C. Jr.; Herne, T. M.; Myles, D. C. *J. Am. Chem. Soc.* **1997**, *119*, 12980.
2.   We have found one other system where this type of immobilization is used - Katz, E.; Schlereth, D. D.; Schmidt, H.-L. Olsthoorn, A. J. J. *J. Electroanal. Chem.* **1994**, *368*, 165.
3.   Ulman, A. *An Introduction to Ultrathin Organic Films: From Langmuir-Blodgett to Self-Assembly*; Academic Press: New York, NY, 1991.
4.   Ulman, A; Eilers, J. E.; Tillman, N. *Langmuir* **1989**, *5*, 1147.
5.   Folkers, J. P.; Laibinis, P. E.; Whitesides, G. M. *Langmuir* **1992**, *8*, 1330.
6.   Ulman, A.; Evans, S. D.; Shnidman, Y.; Sharma, R.; Eilers, J. E.; Chang, J. C. *J. Am. Chem. Soc.* **1991**, *113*, 1499.
7.   Nuzzo, R. G.; Zegarski, B. R.; Dubois, L. H. *J. Am. Chem. Soc.* **1987**, *109*, 733.
8.   Zhang, L.; Lu, T.; Gokel, G. W.; Kaifer, A. E. *Langmuir* **1993**, *9*, 786.
9.   Whitesides, G. M.; Laibinis, P. E. *Langmuir* **1990**, *6*, 87.

198

10. Xia, Y.; Mrksich, M.; Kim, E.; Whitesides, G. M. *J. Am. Chem. Soc.* **1995**, *117*, 9576.
11. Wilbur, J. L.; Kumar, A.; Kim, E.; Whitesides, G. *Adv. Mater.* **1994**, *6*, 600.
12. Abbott, N.; Kumar, A.; Whitesides, G. M. *Chem. Mater.* **1994**, *6*, 596.
13. Motesharei, K.; Myles, D. C. *J. Am. Chem. Soc.* **1997**, *119*, 6674.
14. Buijd, J.; Britt, D. W.; Hlady, V. *Langmuir* **1998**, *14*, 335.
15. Motesharei, K.; Myles, D. C. *J. Am. Chem. Soc.* **1994**, *116*, 7413.
16. Legget, G. J.; Roberts, C. J.; Williams, P. M.; Davies, M. C.; Jackson, D. E.; Terndler, S. J. B. *Langmuir*, **1993**, *9*, 2356.
17. Mrksich, M.; Whitesides, G. M. *Tibtech*, **1995**, *13*, 228.
18. Singvi, R.; Kumar, A.; Lopez, G. P.; Bhatia, S. K.; Hickman, J. J.; Ligler, F. S. *J. Am. Chem. Soc.* **1992**, *114*, 4432.
19. Ebara, Y.; Okahata, Y. *Langmuir* **1993**, *9*, 574.
20. Miyasaka, T.; Koichi, K.; Watanabe, T. *Chem Lett.* **1990**, 627.
21. Willner, I.; Riklin, A.; Shoham, B.; Rivenzon, D.; Katz, E. *Adv. Mater.* **1993**, *5*, 912.
22. Willner, I.; Riklin, A. *Anal. Chem.* **1994**, *66*, 1535.
23. Collinson, M.; Bowden, E. F. *Langmuir* **1992**, *8*, 1247.
24. Prime, K. l.; Whitesides, G. M. *J. Am. Chem. Soc.* **1993**, *115*, 10714.
25. Prof. Mark Porter (Ames, Iowa) has reported a similar aldehyde-terminated monolayer. Personal communication and Electrochemical Society Meeting, Chicago, IL, October, 1995.
26. Leonard, N. J.; Johnson, C. R. *J. Org. Chem*, **1962**, *27*, 282.
27. Konno, K.; Hashimoto, K.; Shirahama, H.; Matsumoto, T. *Tetrahedron Lett.* **1986**, *27*, 3865.
28. Lin-Vien, D.; Colthup, N.B.; Fateley, W. G.; Grasselli, J. G. eds. *The Handbook of Infrared and Raman Characteristic Frequencies of Organic Molecules*, Academic Press: NY, 1991, pp. 122.
29. Bain, C. D.; Evall, J.; Whitesides, G. M. *J. Am. Chem. Soc.* **1989**, *111*, 7155.
30. Porter, M. D.; Bright, T. B.; Allara, D. L.; Chidsey, C. E. D. *J. Am. Chem. Soc.* **1987**, *109*, 3559.
31. Bain, C. D.; Whitesides, G. M., *J. Phys. Chem.* **1989**, *93*, 1670.
32. March, J. *Advanced Organic Chemistry; Reactions, Mechanisms, and Structure*, John Wiley & Sons: New York, 1992, p. 884.
33. Boyd, D. R.; Coulter, P. B.; Sharma, N. D. *Tetrahedrom Lett.* **1985**, *26*, 1673.
34. Boyer, J. H. *Chem Rev.* **1980**, *80*, 495.
35. Shainyan, B. A.; Mirskova, A. N. *Russ. Chem. Rev.* **1979**, *48*, 107.
36. Beschke, H. *Aldrichimica Acta* **1981**, *14*, 13.
37. Tarlov, M. J. *Langmuir* **1992**, *8*, 80

Chapter 15

Nanometer Scale Fabrication of Self-Assembled Monolayers: Nanoshaving and Nanografting

Gang-yu Liu and Song Xu

Department of Chemistry, Wayne State University, Detroit, MI 48202

Two fabrication techniques, nanoshaving and nanografting, have been developed to produce nanopatterns of self-assembled monolayers. Nanoshaving creates negative nanofeatures by displacing self-assembled adsorbates using an atomic force microscopy tip under high imaging forces. Both positive and negative patterns can be produced via nanografting, which is a new method recently developed by our research group. Nanografting combines adsorbate displacement and self-assembly of alkanethiols on gold. This chapter discusses potential applications and important technical issues involved in both methods such as the determination of the optimal fabrication forces, and the influence of the environment on the precision of nanofeatures. Compared with other microfabrication methods, these two techniques allow more precise control of size and geometry. Resolution better than 1 nm can be routinely obtained. In nanografting, multiple component features are fabricated and nanostructures can be quickly changed, modified, and characterized *in situ*. These advantages make both methods very useful for the fabrication and characterization of prototypical nanoelectronic devices.

Microfabrication of self-assembled monolayers (SAMs) has attracted tremendous attention because SAMs have becoming promising candidates as resists used for pattern transfer (1-3). Patterns with dimension of 0.1 μm or larger have been fabricated within a SAM using photolithography (4-8), and recently, via microcontact printing (2,9-17), microwriting (2,9,16) and micromachining (16). Argon ion or electron beam lithography can produce patterns as small as tens of nanometers (18-20). Another approach of constructing nanopatterns of SAMs is to use mixed components in solution (21-25). The distribution of the size and geometry of these domains is determined by the interplay of the kinetics and thermodynamics of the self-assembly process (21-25). Creating nanometer or molecular sized patterns on SAMs in a controlled manner is a remaining. Attempts to accomplish these challenges have triggered many reports of scanning probe microscopy-(SPM) based fabrications, resulting in an emerging field of scanning probe lithography (SPL) (26). Thus far, almost all nanopatterns of SAMs produced via SPL are negative patterns, e.g. holes or trenches created by selective removal of adsorbates (26). The removal of SAMs can be achieved either mechanically via an atomic force microscopy (AFM) tip under local high pressure and shear force (27-35), or with a scanning tunneling microscopy (STM) tip during voltage pulsing

(36,37). In this chapter, we discuss the fabrication of both negative and positive nanopatterns of SAMs using AFM based lithography. Technical issues involved in nanofabrication will be addressed such as determination of optimized displacement force, and the influence of the SAM structure and surrounding environments on the precision of the nanofeatures.

Principles of Nanoshaving and Nanografting

Basic procedures Positive and negative nanopatterns can be fabricated using nanoshaving and nanografting, respectively. The three basic steps of both techniques are illustrated in Figure 1. The first step is surface structure characterization via AFM in which SAM resists are imaged under a very low force. Fabrication locations are normally selected at areas with flat surface morphology, e.g. a Au(111) plateau area. The second step is fabrication. In nanoshaving, fabrication is accomplished by exerting a high local pressure using an AFM tip. This pressure results in a high shear force during the scan, which causes the displacement of SAM adsorbates. In this chapter, we differentiate between nanoshaving and wearing (38). In nanoshaving, SAM adsorbates are displaced by scanning an AFM tip under a load higher than the displacement threshold (27,28,30). Holes and trenches can be fabricated with one or several scans. Wearing is normally referred to as detaching of adsorbates by repeatedly scanning an AFM tip under a smaller load than the threshold (27-29). In wearing, molecules are gradually moved away from the edges of defect sites. During AFM scans, both processes may occur simultaneously although shaving is the dominant process under high load. Nanografting is a new procedure recently developed by our research group (27). In nanografting, thiol molecules within a SAM are displaced in a solution containing reactive adsorbates. These active molecules grow on the newly exposed substrate as an AFM tip plows through the matrix SAM (27). Depending upon the relative chain length between the active component and the matrix SAM, positive or negative patterns may be created. The fabricated nanofeatures are characterized in the third step after reducing the load.

Nanoshaving   **Nanografting**

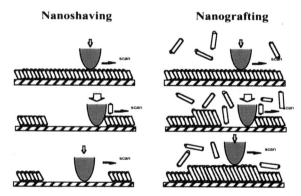

Figure 1. Schematic diagram of the three basic steps for nanoshaving and nanografting.

Determination of fabrication forces The fabrication force is one of the most important parameters for nanoshaving and nanografting. Overloading the force can cause plastic deformation or even displacement of the underlying substrate (39-41). On the other hand, if the force is too low, adsorbates cannot be removed completely and quickly. *The optimized fabrication force, based on our experience, should be set slightly above the displacement threshold.* Since the fabrication force threshold varies with the geometry of AFM tips, the structure of the matrix SAMs, and the fabrication environment (27,28,38), it is recommended that the threshold should be determined *in situ* for each individual experiment before attempting fabrication.

Using high resolution AFM imaging, the force threshold is determined by monitoring the changes in surface structure as a function of increasing load. Under low imaging forces, topographic images reveal the surface structures of the SAMs. Figure 2 shows an ordered n-alkane thiol monolayer on gold and a disordered siloxane monolayer on mica. As the load increases, the images remain unchanged at first, and become more and more distorted at higher forces. A continuous increase of the imaging force results in a transition in the AFM image from the lattice of the SAM to that of the substrate (Figure 2). The load at which the transition occurs is referred to as the displacement threshold.

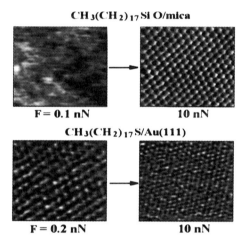

Figure 2. Topographic images (50 x 50 Å²) of octadecanethiol (C₁₈S/Au(111) and octadecyltriethoxysilane (OTE/mica) SAMs taken in 2-butanol. The C₁₈SH layer exhibits a commensurate ($\sqrt{3} \times \sqrt{3}$)R30° structure on Au(111), while the OTE/mica does not show long range orders. At a load larger than 9.7 nN, the corresponding images suddenly change to show the periodicity of Au(111) and mica (0001). The threshold pressure for this experiment is estimated to be 0.4 Gpa using the Hertzian model (30-32,38). The tip radii are ca. 75 Å, as calibrated by imaging the Au(111) single atomic steps.

Nanoshaving

In order to produce high resolution patterns during nanoshaving, ideal conditions include molecule-by-molecule displacement, immediate removal of the displaced adsorbate, and slow readsorption rates. The fate of the displaced adsorbates depends upon the structure of SAMs and the fabrication environment. Thiols/gold and siloxanes/mica represent two important categories of SAMs. As illustrated in Figure 3, thiol molecules chemisorb on Au(111) to form a close-packed structure (42-50). There is no cross-linking among the thiol chains (42). The chemisorption energy is ~ 40

202

Kcal/mol, and the van der Waals energy per CH_2 group is ~ 2 Kcal/mol (38). Under high local pressure and shear force, thiol molecules can be displaced from their adsorption sites, but may still remain weakly bound to gold in a nearby location. Therefore, without solvation or other external forces to remove these displaced thiols, they are likely to move laterally and become chemisorbed on gold after the local pressure is removed (30-32). Figure 4 depicts the results of nanoshaving of $C_{18}S/Au(111)$ performed under ambient laboratory conditions, in 2-butanol and in water. Under ambient laboratory conditions, the displacement of thiols is generally reversible (30-32). Partial reversibility is also observed if permanent wearing occurs (Figure 2). In water, the displacement is also reversible due to the poor solubility of long chain thiols in water. In 2-butanol or other organic solvents in which $C_{18}SH$ exhibits sufficient solubility, the displacement becomes irreversible allowing the best fabrications to be achieved.

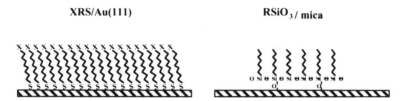

Figure 3. Schematic, cross-sectional diagram of two representative SAMs: thiols/gold and siloxanes/mica.

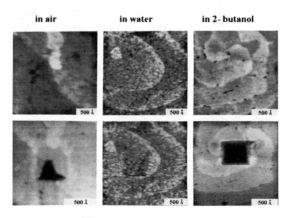

Figure 4. A 1600 x 1600 Å^2 topographic images of $C_{18}S/Au(111)$ before (top raw) and after (bottom raw) nanoshaving under three different environments. The central 500 x 500 Å^2 area is fabricated in each case, and borders of the shaved areas can be clearly identified from the structural discontinuity. The cantilevers used for all experiments are sharpened microlevers (0.1 N/m) from Park Scientific Instruments.

In contrast to thiol SAMs on gold, siloxane layers on mica do not exhibit long range order (28,51-54). The inter-chain van der Waals interactions for siloxane SAMs are weaker than that in the corresponding thiol SAMs, due to the non-close packed structure of the siloxane molecules. There are very few covalent bonds between the siloxane molecules and the mica substrate, due to the low density of dangling hydroxyl groups on the mica surface (28,54). The stability of siloxane SAMs arises from the cross-linking of the chains through the Si-O network, which are formed by baking the adsorbate layer at 120°C for ça. 2 hrs (28,54). Due to the low reactivity between the displaced siloxane molecules and mica, the displacement of siloxane SAMs at room temperature is always irreversible (28,29). Siloxane molecules are either removed to a distance further away from the fabrication sites during the scan, or else they become attached to the AFM tip (55). Figure 5 shows the nanoshaving of an octadecyltriethoxysilane (OTE) on mica in air, 2-butanol and water.

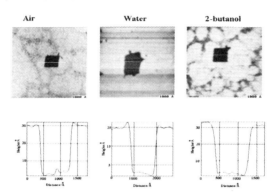

Figure 5. A 5000 x 5000 Å^2 topographic image of OTE/mica. The OTE molecules in the central area (100 x 100 Å^2) are removed using nanoshaving in air, 2-butanol and water respectively. The corresponding cursor profiles indicate that the thickness of the layer is 25 Å, and the displacement is complete under all three conditions.

Nanografting

Nanografting can create both positive and negative patterns depending upon the relative chain length between the new and matrix adsorbates. Most of our nanografting experiments have been conducted using a thiol SAM matrix in a 2-butanol solution containing thiol(s) (27). Figure 6 illustrates an example in which two $C_{18}S$ nanoislands are fabricated in a matrix monolayer, $C_{10}S/Au(111)$. Edge resolution higher than 1 nm can be routinely obtained (27). It is important to point out that the newly grafted $C_{18}S$ nanoislands not only have an ordered and close-packed $(\sqrt{3} x \sqrt{3})R30°$ lattice, but also are free of defects such as pin holes or uncovered areas (27). The absence of pin hole defects is critical for faithful pattern transfer when patterned SAMs are used as masks (1-3).

Taking advantage of the high mechanical stability of our home-constructed AFM scanning head, we are able to replace the thiol solution without causing a significant lateral drift between the tip and the surface. This advantage allows multiple patterns to be fabricated with any desired arrangement and composition of thiols. In the example shown in Figure 7, two patterns, a 250 x 600 Å^2 $C_{22}S$ and a 200 x 600 Å^2 $C_{18}S$ island, are grafted side-by-side and inlaid into the $C_{10}S$ matrix layer. The observed heights (cursor plot in Fig. 7) and high resolution images indicate that the chains are closely-packed within both nanoislands. The ability to fabricate multiple component patterns opens possibilities for producing nanochips and sensors with orthogonal detection capabilities (56).

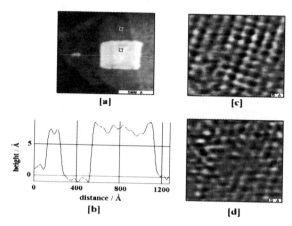

Figure 6. (a) Two $C_{18}S$ nanoislands (30 x 50 and 500 x 500 Å^2) are fabricated in the matrix of a $C_{10}S$ monolayer via nanografting. Single atomic steps of Au(111) are used as a landmark. As shown in the cursor profile in (b), the $C_{18}S$ islands are 8.8 Å higher than the surrounding $C_{10}S$ monolayer, consistent with the theoretical value for crystalline-phased SAMs. Higher resolution images (c and d) acquired from the matrix (white square in a) and the grafted islands (black square in a) areas have indicated the $(\sqrt{3} \times \sqrt{3})R30°$ structure in both $C_{10}S$ and $C_{18}S$ areas.

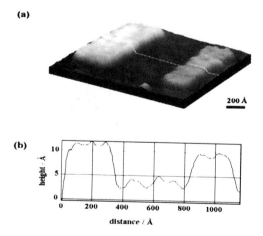

Figure 7. (a) Fabrication of multicomponent patterns using nanografting. The lateral dimensions of the $C_{22}S$ and $C_{18}S$ islands are 250 x 600 Å^2 and 200 x 600 Å^2. (b) The corresponding cursor profile shows that the $C_{22}S$ and $C_{18}S$ islands are 7.5 ± 1.0 Å and 12.0 ± 1.5 Å above the matrix monolayer respectively.

In the development of prototypical nanoelectronic devices, it is important to be able to systematically change or modify the geometry of the nanostructures in order to optimize performance. Nanografting enables us to quickly change and/or modify the fabricated patterns *in situ* (27). An example illustrating this concept is shown in Figure 8. First, two parallel $C_{18}S$ nanolines were fabricated, and then the distance between them was increased following the procedure shown in Figure 8. In contrast to lithography or other microfabrication methods, modification of patterns via nanografting does not require changes of the mask or repetition of the entire fabrication experiment. This advantage provides a unique opportunity for researchers to study size-dependent properties unambiguously, because other experimental conditions can be held constant .

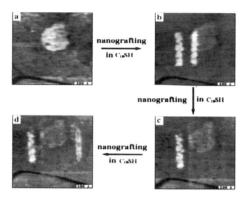

Figure 8. Changing the nanostructures *in situ*. (a) AFM image of a $C_{10}S/Au(111)$ before fabrication. The bright area is a single atomic Au(111) step which is used as a landmark for the fabrication process. (b) Using nanografting(illustrated in Figure 1), two parallel $C_{18}S$ nanolines are fabricated with the dimensions of 100 x 500 $Å^2$, and a separation of 200 Å. Afterwards, the $C_{18}SH$ solution is replaced by $C_{10}SH$ solution by flowing excess amount of $C_{10}SH$ solution through the AFM liquid cell. (c) Erasure of one line by scanning the nanoline area under high imaging force in a $C_{10}SH$ solution. The original pattern is completely erased, and the frame of the scanned areas can be very vaguely identified due to the structural discontinuity at the border. The $C_{10}SH$ solution is then replaced by $C_{18}SH$ solution. (d) Refabrication of the second line by nanografting in the $C_{18}SH$ solution again. The interline spacing now is increased to 60 nm.

Summary

Two nanofabrication methods of SAMs, nanoshaving and nanografting, are discussed in this chapter. Nanoshaving relies on high local pressure and shear force to displace SAM adsorbates, and nanografting is achieved by simultaneous displacement of the matrix layer and adsorption of a new SAM. Compared with other methods used to fabricate microstructures of SAMs, nanoshaving and nanografting offer the advantage of high spatial resolution. Edge resolution of 1 nm are routinely obtained and molecular precision can be achieved with an ultrasharp tip. In addition, nanostructures can be characterized with molecular resolution *in situ* using the same AFM tip.

Nanopatterns with different thiol components can be fabricated with desired arrangements via nanografting, which provides a unique opportunity for fabrication of orthogonal sensors (56). Once a nanografting experiment is set up, one can quickly change and/or modify the fabricated patterns *in situ* without changing the mask or repeating the entire fabrication procedure. There capabilities allow systematic studies of size-dependent properties for nanostructures. No doubt, both methods, especially nanografting, are still in their infancies, and further studies are necessary to develop them into a generic technique for nanofabrication. In combination with pattern transfer protocols, such as selective etching (16,57,58) or deposition (10,59-62), these methods may be used to form novel nanoelectronic devices for research purposes.

Acknowledgments

We thank Professor Paul E. Laibinis at MIT for the $CH_3(CH_2)_{21}SH$ compound. We appreciate many helpful discussions with Professor Bill Hase, Jayne Garno, Scott Miller, Vijay Jain and Kapila Wadu-Mesthrige. SX thanks WSU-IMR for a graduate research fellowship. GYL gratefully acknowledges the Camille and Henry Dreyfus Foundation for a New Faculty Award, and the Arnold and Mabel Beckman Foundation for a Young Investigator Award. This work is also supported by Wayne State University, the Institute for Manufacturing Research, and National Science Foundation Grant CHE- 9733400.

References

(1)   Bishop, A.; Nuzzo, R. G. *Current Opinion in Colloid Interface Sci.*, **1996**, *1*, 127.

(2)   Kumar, A.; Abbott, N. L.; Kim., E.; Biebuyck, H. A.; Whitesides, G. M. *Acc. Chem. Res.* **1995**, *28*, 219.

(3)   Bumm, L. A.; Arnold, J. J.; Cygan, M. T.; Dunbar, T. D.; Burgin, T. P.; Jones, L.; Allara, D. L.; Tour, J. M.; Weiss, P. S. *Science* **1996**, *271*, 1705.

(4)   Huang, J. Y.; Dahlgren, D. A.; Hemminger, J. C. *Langmuir* **1994**, *10*, 626.

(5)   Tarlov, M. J.; Burgess, D. R. F.; Gillen, G. *J. Am. Chem. Soc.* **1993**, *115*, 5305.

(6)   Wollman, E. W.; Kang, D.; Frisbie, C. D.; Lorkovic , I. M.; Wrighton, M. S. *J. Am. Chem. Soc.* **1996**, *116*, 4395.

(7)   Calvert, J. M. "lithographically Patterned Self-Assembled Films" Ed. Ulman, A. D. *Thin Films*, Academic Press, Boston, **1995**, *20*, 109.

(8)   Fodor, S. P. A.; Read, J. L.; Pirrung, M. C.; Stryer, L.; Lu, A. T.; Solas, D. *Science*, **1991**, *251*, 767.

(9)   Abbott, N. L.; Kumar, A.; Whitesides, G. M. *Chem. Mater.* **1994**, *6*, 596.

(10)   Sondag-Huethorst, J. A. M.; Van Helleputte, H. R. J.; Fokkink, L. G. J. *Appl. Phys. Lett.* **1994**, *64*, 285.

(11)   Tiberio, R. C.; Craighead, H. G.; Lercel, M.; Lau, T.; Sheen, C. W.; Allara, D. L. *Appl. Phys. Lett.* **1993**, *62*, 476.

(12)   Berggren, K. K.; Bard, A.; Wilbur, J. L.; Gillaspy, J. D.; Helg, A. G.; McClelland, J. J.; Rolston, S. L.; Phillips, W. D.; Prentiss, M.; Whitesides, G. M. *Science* **1995**, *269*, 1255.

(13) Kumar, A.; Abbott, N. L.; Kim, E.; Biebuyck, H. A.; Whitesides, G. M. *Acc. Chem. Res.* **1995**, *28*, 219.
(14) Kumar, A.; Whitesides, G. M. *Science* **1994**, *263*, 60.
(15) Xia, Y.; Whitesides, G. M. *J. Am. Chem. Soc.* **1995**, *117*, 3274.
(16) Kumar, A.; Biebuyck, M. A.; Abbott, N. L.; Whitesides, G. M. *J. Am. Chem. Soc.* **1992**, *114*, 9188.
(17) Jackman, R. J.; Wilbur, J. L.; Whitesides, G. M. *Science* **1995**, *269*, 664.
(18) Sondag-Huethorst, J. A. M.; Van Helleputte, H. R. J.; Fokkink, L. G. J. *Appl. Phys. Lett.* **1994**, *64*, 285.
(19) Tiberio, R. C.; Craighead, H. G.; Lercel, M.; Lau, T.; Sheen, C. W.; Allara, D. L. *Appl. Phys. Lett.* **1993**, *62*, 476.
(20) Berggren, K. K. et al, *Science* **1995**, *269*, 1255.
(21) Bard, A. J.; Abruna, H. D.; Chidsey, C. E. D; Faulkner, L. R.; Feldberg, S. W.; Itaya, K.; Majda, M.; Melroy, O.; Murray, R. W.; Porter, M. D.; Suriaga, M. P.; White, H. S. *J. Phys. Chem.* **1993**, *28*, 7147.
(22) Chidsey, C. E. D. *Science*, **1991**, *251*, 919.
(23) Chidsey, C. E. D.; Bartozzi, C. R.; Putvinski, T. M.; Majsie, T. M. *J. Am. Chem. Soc.*, **1990**, *112*, 4301.
(24) Hayes, W. A.; Kim, H.; Yue, X.; Perry, S. S.; Shannon, C. *Langmuir,* **1997**, *13*, 2511.
(25) Bilewicz, R.; Sawaguchi, T.; Chamberlain, R. V.; Majda, M. *Langmuir*, **1995**, *11*, 2256.
(26) Nyffenegger, R. M.; Penner, R. M. *Chem. Reviews* **1997**, *4*, 1195.
(27) Xu, S.; Liu, G. Y. *Langmuir,* **1997**, *13*, 127.
(28) Xiao, X. D.;. Liu, G. Y.; Charych, D. H.; Salmeron, M. *Langmuir*, **1995**, *11*, 1600.
(29) Kiridena, W.; Jain, V.; Kuo, P. K.; Liu, G. Y. Surf. Interf. Analysis, **1997**, *25*, 383.
(30) Liu, G.-Y.; Salmeron, M. B. *Langmuir* **1994**, *10*, 367.
(31) Liu, G.-Y.; Fenter, P.; Eisenberger P.; Chidsey, C. E. D.; Ogletree, D. F.; Salmeron, M. *J. Chem. Phys.* **1994**, *101*, 4301.
(32) Salmeron, M. B.; Liu, G.-Y.; Ogletree, D. F. *Forces in Scanning Probe Methods* eds. Gutherodt, H. J.; Anselmetti, D.; Meyer, E. NATO ASI Series, E. Kluwer Academic Publisher the Neitherlands, **1995**, 593.
(33) Jung, T. A.; Moser, A.; Hug, H. J.; Brodbeck, D; Hofer, R.; Hidber, H. R.; Schwartzm U. D. *Ultramicroscopy* **1992**, *42-44,* 1446 North - Holland.
(34) Garnaes, J.; Bjonholm, T.; Zasadzinski, J. A. N. *J. Vac. Sci, Technol B.,* **1994**, *12*, 1839.
(35) Chi, L. F.; Eng, L. M.; Graf, K.; Fuchs, H. *Langmuir*, **1992**, *8*, 2255.
(36) Ross, C. B.; Sun , L.; Crooks, R. M.; *Langmuir* **1993**, *9*, 632.
(37) Schoer, J. K.; Ross, C. B.; Crooks, R. M.; Corbitt, T. S.; Hampden-Smith, M. J. *Langmuir* **1994**, *10*, 615.
(38) Carpick, R. W.; Salmeron, M. *Chem. Reviews,* **1997**, *4,* 1163.
(39) Salmeron, M. B.; Folch, A.; Neubauer, G.; Tomitori, M.; Ogletree, D. F.; Kolbe, W. *Langmuir* **1992**, *8*, 2832.
(40) Ju, J.; Xiao, X. D.; Ogletree, D. F.; Salmeron, M. *Surf. Sci.* **1995**, *327*, 358.
(41) Salmeron, M.; Neubauer, G.; Folch, A.; Tomitori, M.; Ogletree, D. F.; Sautet, P. *Langmuir*, **1993**, *9*, 3600.
(42) Poirier, G. E. *Langmuir*, *13*, 2019.
(43) Schonenberger, C.; Sondag-Hüthorst, J. A. M.; Jorritsima, J.; Fokkink, L. J. G. *Langmuir*, **1994**, *10*, 611.
(44) McDermott, C. A.; McDermott, M. T.; Green, J. B.; Porter, M. D. *J. Phy. Chem.* **1995**, *99*, 13257.

208

(45) Chidsey, C. E. D.; Liu, G. Y.; Rowntree, P.; Scoles, G. *J. Chem. Phys.* **1989**, *91(7)*, 4421.

(46) Chidsey, C. E. D., Liu, G-Y., Putvinski, T. M.; Scoles, G. *J. Chem. Phys.*, **1991**, *94(12)*, 8493.

(47) Chidsey, C. E. D.; Liu, G-Y.; Scoles, G. *J. Chem. Phys*, **1993**, *98(5)*, 4234.

(48) Chidsey, C. E. D.; Eisenberger, P.; Fenter, P. Li; J.; Liang, K. S.; Liu, G-Y.; Scoles, G. *J. Chem. Phys.*, **1993**, *99(1)*, 744.

(49) Fenter, P.; Eberhardt, A.; Eisenberger, P. *Science*, **1994**, *266*, 1216.

(50) Fenter, P.; Eisenberger, P.; Liang, K. S. *Phys. Rev. Lett*, **1993**, *70*, 2447.

(51) Ulman, A. **1991**, Academic Press, San Diego, CA.

(52) Schwartz, D. K.; Steinberg, S.; Israelachvilli, J.; Zasadzinski, J. A. N. *Phy. Rev. Lett.* **1992**, *69*, 3354.

(53) Maoz, R.; Yam, R.; Berkovic, G.; Sagiv, J. *Organic Thin Films and Surfaces: Directions for the Nineties*, ed. Ulman A., Academic Press, Boston, *Thin Films*, **1995**, *20*, 41.

(54) Kessel, C. R.; Granick, S. *Langmuir*, **1991**, *7*, 532.

(55) Jaschke, M.; Butt, H.-J. *Langmuir*, **1995**, *11*, 1061.

(56) Thomas, R. C.; Yang, H. C.; Dirubio, C. R.; Ricco, A. J.; Crooks, R. M. *Langmuir*, **1996**, *12*, 2239.

(57) Xia, Y.; Zhao, X. M.; Kim, E.; Whitesides, G. M. *Chem. Mater.* **1995**, *7*, 2332.

(58) Kim, E.; Kumar, A.; Whitesides, G. M. *J. Electrochem. Soc.* **1995**, *142*, 628.

(59) Jeon, N. L.; Nuzzo, R. G.; Xia, Y.; Mrksich, M.; Whitesides, G. M. *Langmuir* **1995**, *11*, 3024.

(60) Potochnik, S. J.; Pehrsson, P. E.; Hsu, D. S. Y.; Calvert, J. M. *Langmuir*, **1995**, *11*, 1841.

(61) Hampden-Smith, M. J.; Kodas, T. T. *Chem. Vap. Deposition*, **1995**, *1*, 8.

(62) Jeon, N. L.; Clem, P. G.; Payne, D. A.; Nazzo, R. G. *Langmuir*, **1996**, *12*, 5350.

INDEXES

Author Index

Subject Index

A